Ethnobiology

Series editor
Robert Voeks, Center for Remote Sensing & California State University, Fullerton, California, USA
John Richard Stepp, Department of Anthropology, University of Florida, Gainesville, Florida, USA

Ethnobiology

Ethnobiology is the study of the dynamic relationship between plants, animals, people, and the environment. Academic and applied interests include ethnobotany, ethnozoology, linguistics, paleoethnobotany, zooarchaeology, ethnoecology, and many others. The field lies at a dynamic intersection between the social and biological sciences. The major contribution from the biological sciences has come from Economic botany, which has a rich historical and scientific tradition. Indeed, the objectives of the colonial enterprise were as much about the quest for "green gold" –herbal medicines, spices, novel cultivars, and others—as it was for precious metals and sources of labor. The view that ethnobiology concerns mostly the discovery of new and useful biota extended into the 20th century, fortified considerably by work of Richard Evans Schultes and his students. The social sciences have contributed to the field in both descriptive studies but also within quantitative approaches in cognitive anthropology that have led to general principles within ethnobiological classification. Ethnobiological research in recent years has focused increasingly on problem solving and hypothesis testing by means of qualitative and especially quantitative methods. It seeks to understand how culturally relevant biotas are cognitively categorized, ranked, named, and assigned meaning. It investigates the complex strategies employed by traditional societies to manage plant and animal taxa, communities, and landscapes. It explores the degree to which local ecological knowledge promotes or undermines resource conservation, and contributes to the solution of global challenges, such as community health, nutrition, and cultural heritage. It investigates the economic value and environmental sustainability to local communities of non-timber forest products, as well as the strategies through which individual ecological knowledge and practices encourage resilience to change—modernization, climate change, and many others. Most importantly, contemporary ethnobiological research is grounded in respect for all cultures, embracing the principles of prior informed consent, benefit sharing, and general mindfulness.

More information about this series at http://www.springer.com/series/11551

Takuro Furusawa

Living with Biodiversity in an Island Ecosystem

Cultural Adaptation in the Solomon Islands

Springer

Takuro Furusawa
Department of Ecology and Environment,
Division of Southeast Asian Area Studies,
Graduate School of Asian and African Area Studies
Kyoto University
Kyoto, Japan

ISSN 2365-7553 ISSN 2365-7561 (electronic)
Ethnobiology
ISBN 978-981-13-5725-1 ISBN 978-981-287-904-2 (eBook)
DOI 10.1007/978-981-287-904-2

Springer Singapore Heidelberg New York Dordrecht London

Softcover re-print of the Hardcover 1st edition 2016

Printed on acid-free paper

Springer Science+Business Media Singapore Pte Ltd. is part of Springer Science+Business Media (www.springer.com)

A local forest expert demonstrates how to drink tree sap flowing from a freshly cut vine (*Uncaria appendiculata*, locally known as *vinetuŋu malivi*)

Preface

Encounter in the Solomon Islands

> A variety of trees grow on the riverbank and many birds come to use those trees. Droppings of these birds fall into river and streams; the roots of the mangrove at the river's mouth catch the droppings. These droppings enrich the mangrove soils and also provide a preferred food of fish. Thus, several kinds of fish come to the mangrove to nest and lay their eggs. Birds have several other roles including contributing to people. If a bird roosts on a tree in a garden, her droppings enrich the soil for crops. She may eat nuts and fly away, but when she lands she only eats the fruity flesh and drops living seeds in a remote area. New nut trees grow from those seeds and we people eat them.

About 4 months after I started my first fieldwork in Olive village, Roviana Lagoon, Western Province, Solomon Islands, I started a conversation with an impressive informant. Her story, as described above, showed a rich and deep knowledge of the local ecosystem. Hearing it encouraged me to conduct additional research on the interactions between people and the natural world in this area. Her story supported my hypothesis that the people of Roviana recognized the existence of ecological webs of human–plant–animal interaction and that this recognition played an important role in subsistence, resource use, and the conservation of natural resources, as well as land management.

This village, located on an island covered by dense tropical rain forest, provided a home for villagers who collected forest resources for their livelihood and cleared small areas of land for the production of root crops. The island displayed a rich level of biodiversity. Even though only a few land mammals thrive on such an island, a wide variety of abundant birds filled the forest, settlements, which have gardens with a variety of bird songs. The village faced a large coral lagoon, where the villagers angled or netted fishes. The informant was a woman in her mid-50s, an "ordinary" villager, i.e., neither a customary leader nor a healer, who had only completed 4 years of primary school education. However, her knowledge of the forest sounded comprehensive and ecologically sound. I was sometimes surprised by the depth of her knowledge, especially as it related to how different creatures—including humans—interacted within the environment where she lived. She insisted that she had developed this knowledge through her own experiences rather than learning from others. She continued her story:

> Since commercial logging started in our territory, the birds have fewer places to live. As a result, fewer birds nest in the trees and enrich our soils with their droppings. Seed dispersal has also decreased and fewer fishes come to the mangroves. Listen. In the past, birds helped the trees grow and we had a forest. We people are thankful for the birds that scatter seeds, allowing new trees to grow, because we may receive royalty money from the logging company. However, logging robs the birds of their places to live and make their nests.

The development of the rural area of Roviana accelerated after the logging operations started. This caused the forest to deteriorate and changed the lifestyles of the villagers. In addition to changes in forest management, Olive village experienced rapid changes under the developing market economy that coincided with population growth and urbanization. Unfortunately, the adverse pressure that development created on the forest came from multiple sources; both internal and external factors were involved. That is, the people themselves chose to accept the logging operation, to expand cash cropping by clearing the forest, and to sell forest resources with the goal of improving the quality of their lives, even though these behaviors were partly driven by unavoidable globalization and local population growth. I felt a little confused and did not initially understand why people who had such rich ecological knowledge were simultaneously willing to exploit forest resources.

Purpose of the Study

I primarily intend to document the subsistence lifestyle and lifeways of the people of Roviana in relation to the natural world, their indigenous ecological knowledge, and related recent changes in the Roviana Lagoon, Solomon Islands, using a transdisciplinary approach that integrated ecology, ethnography, agriculture, remote sensing, and statistics. My fieldwork, which started in 2001, serves as the basis for this study. This research provides an interesting case study of ecological and cultural interaction between humans and a landscape with a high level of biodiversity, and is designed to analyze the changes occurring in this society and ecosystem. This study analyzes how humans and biodiversity interact and also how humans and the ecosystem react to the socioeconomic changes occurring in the Solomon Islands. This study will provide informative suggestions that are designed to address a series of general questions, such as, "What kinds of factors have determined whether current human actions are sustainable or will result in a collapse of biocultural diversity in the Solomon Islands?" "How do Solomon Islanders recognize nature and biodiversity conservation in traditional ways or under socioeconomic changes?" and "How can the human societies of the Solomon Islands achieve that harmonization under changing socioeconomic conditions?" Therefore, my efforts were spent on collecting (1) human-related ecological data that could be used to analyze how people produced and consumed both food and other resources in a tropical rain forest ecosystem, (2) botanical data revealing what kinds of resources they used, (3) data on biodiversity showing how this type of production, consumption, and resource use depended on the natural world and how that world was positively or negatively affected by these human activities, (4) psychological (or traditional anthropological and sociological knowledge) data designed to reveal what recognition underlies such activities, (5) socioeconomic data showing how recent changes affected the interactions between humans and the natural world around them, and (6) qualitative data related to what and how social institutions and cultural factors worked to support resource management and conservation. Additionally, human health and biological evolution were also studied to provide a background related to how humans adapt to change, although these data were mentioned only in passing in this book. Overall, this book provides a multidisciplinary and thorough study of the interaction between humans and the natural world on a relatively small spatial scale in the Pacific and islands of Southeast Asia.

Human ecology is the study of human adaptation and adaptability in an ecosystem, both biologically and culturally. A number of studies have attempted to understand how humans adapt to change in subsistence-based societies, i.e., how they produce and consume foods through foraging and agriculture. However, I felt that previous studies oversimplified the production system in two ways. First, the ecosystem approach analyzed society at the population level, even though decision making is done at the household or individual level. Second, the interaction between humans and the natural world cannot be analyzed by simply measuring the flow of energy; this type of study requires an analysis of the complex webs of biocultural diversity. Additionally, many studies have concluded that traditional shifting cultivation cycles and indigenous land-management systems are generally sustainable through appropriate land use and food production. However, several recent studies indicated that societies could fail to sustain their current shifting cycles under ongoing demographic and socioeconomic changes. People experience these changes today. In the field, I found that Roviana people used various ecological settings strategically. They adopted different cultivation methods and crops for different settings, and different communities or households adopted different strategies. Exploring such subsistence diversity rather than simply measuring energy flow at the population level now serves as an important new field in the study of human ecology.

Ethnobotany involves not only understanding how local people recognize, use, and manage energy-providing plants but also studying the use of all other plant resources, and can therefore be applied to the study of the human–ecosystem relationships observed here. In this regard, "quantitative ethnobotany" attracted my attention and led to my designing this study with the goal of measuring the cultural importance of each species in different societies and environments. This method also identifies important and useful vegetation or land cover types that local people find useful. Systematically quantifying detailed observations of human behaviors and lifestyles allows me to improve measurements of the usefulness and importance of different species and various types of land use.

Additionally, a new academic field called "folk ecology" or "folk biology" has been defined as the study of cognitive aspects of interactions between humans and the natural world. Dr. Scott Atran and others revealed that Mayan people of Guatemala recognized the existence of ecological human–plant–animal webs and these people's behavior and management of biodiversity was associated with this type of recognition. My intense interest in this study led me to integrate this psychological domain into my analyses of the background of the ways local people use and manage the forest. This view was very important as it related to understanding whether or not local people recognize complicated interactions between different species. Many previous studies had criticized local people for generally exploiting resources for only short-term benefits without considering any undesirable effects of their activities and perhaps without realizing that they were not contributing to the conservation and creation of biodiversity in the way they had hoped. The theories and methodologies of human ecology, quantitative ethnobotany, and the new study of folk ecology can be combined and quantitatively analyzed in a way that quantitatively links different domains of humans and the natural world. This allowed me to simultaneously analyze the ecosystem, human behavior, and human recognition of environmental ecology comprehensively, statistically, and mathematically.

Another motivation was an increasing global interest in social-ecological production landscapes while emphasizing their global importance. Recent debates related to the conservation of biodiversity have encouraged a broader global recognition of the importance of sustainably preserving natural environments even though they are human-modified [e.g., Target 3 in Strategic Plan for Biodiversity 2011–2020 and the Satoyama Initiative in the 10th Meeting of the Convention of the Parties, Convention on Biological Diversity (CBD-COP10), in 2010]. People have long depended on biodiversity and received ecological services used for their own survival and welfare. This did not necessarily mean they destroyed biodiversity or ecosystems, because human modification might contribute to the creation and conservation of biodiversity. For example, the removal of a tree within a forest aids the growth of understory vegetation and pioneer shade-intolerant trees, and these trees provide ecological niches of some kinds of animals. In this way, social–ecological production landscapes may contribute to balancing the conservation of biodiversity with rural development by using resources wisely and by preserving the rights of local people to live on the land.

My research finally suggested the importance of understanding internal diversity. In this context, internal diversity means intra-society, intra-household, or intra-individual diversity. Biological diversity is usually analyzed at various levels such as genetic, species, population, community, and ecosystem. In contrast, the diversity in human society or culture, as an agent that uses biodiversity, has been considered to have been relatively homogeneous in the past. However, some of the villagers in one village may favor natural resource conservation, while others advocate against it, and still others feel neutral. Additionally, even specific individuals may be involved in conflicting helpful, harmful, and neutral actions simultaneously or recognize biocultural (external) diversity. One tree species may serve a useful purpose when a villager needs to use it and seeks it out, but the same species may be considered a weed in agricultural land. Internal diversity thus exists in levels of the ecological unit (individual, household, village), with people having differences in their behavior in different situations, and in different psychological domains. We need to acknowledge that such multilevel types of diversity exist not only in natural environments but also in human societies and such types of diversity should be taken into consideration when attempting to understand their lives, knowledge, and recognition. This viewpoint is important if conservationists and people working for conservation agencies who live and work outside the community want to integrate their planned conservation practices with local people and their traditional environmental knowledge. Through these interdisciplinary data and discussions, I intend to provide useful information for ecologists, anthropologists, conservation practitioners, and rural development planners.

Structure of This Book

The book starts with a review of background information related to the Solomon Islands, human adaptation, and biocultural diversity in Melanesia, as well as socioeconomic changes (Chap. 1). The concept of living with biodiversity and the working hypothesis are explained. Then, Chap. 2 explains the regional geographical context and historical background of the study site.

Chapters 3, 4, 5, 6, 7, 8, and 9 provide findings from various aspects of detailed fieldwork. Chapter 3 describes the vegetation and landscape from both ecological and cultural viewpoints, i.e., the data suggest forests are botanically diverse and the local people recognize that the local landscape reflected this diversity. Chapter 4 shows how people use different ecosystems, specifically the main island and barrier island ecosystems of the study area with their independent geological histories, that are separated by a coral lagoon sea. This chapter addresses a variety of disciplines such as land use, shifting cycles, land tenure, soil nutrients, subsistence production, vegetation, land cover change, and carrying capacity. Chapter 5 employs a quantitative ethnobotanical analysis to explore the kinds of plants the people use for their daily life and compares the local plant and land use patterns of a rural village and an urbanized village. This chapter then addresses how local people depend on biodiversity and how traditional human modifications of the forest contribute to the conservation of biodiversity by integrating the results presented in Chap. 3. These findings illustrate how a local society's use of resources was related to community welfare as well as to environmental preservation.

Chapter 6 explores the psychological aspects of human–environmental–ecological interaction. The people of Roviana identified and acknowledged a variety of interactions that occur between humans, plants, and animals. However, this chapter also shows how the villagers recognized that people may have adverse effects on the ecosystem. These analyses are used to discuss the existence of traditional conservation ethics that are designed to protect biodiversity and may affect it. Chapter 7 then examines the acculturation of individuals to ethnobotanical knowledge in association with the effects of modernization. Cultural consensus analyses and modernity scores suggested that modern Western knowledge is easily integrated into the ethnobotanical knowledge system and is related to the loss of diverse botanical knowledge.

Chapter 8 predicts the future of Roviana, based on comparative research in the urban area of Roviana. Households in Munda town were economically vulnerable and at risk of encountering poverty and other hardships. This result suggested the necessity of taking action to conserve the rural lifestyle and give it a viable future. Chapter 9 provides qualitative information such as customary rules related to forest use and ecosystem management with the goal of exploring the possibility of integrating the ways that local people modify forests into new biodiversity conservation programs. Finally, Chap. 10 summarizes the findings, suggests ways to harmonize the interactions of humans and the natural world of Roviana, and proposes how this system sustainably benefits the people of Roviana.

Kyoto Takuro Furusawa

Acknowledgments

I would like to sincerely thank all the people of Roviana and Solomon Islands for their kind approval, participation, and support of my research. I would especially like to thank Mr. Edwin Huti, Ms. Risa Donga, Mr. Cornelius Murray, Mr. Eki Lee Dagha, Mr. Rex Dagha, Ms. Rosy Dagha, Mr. Rupas Murray, Mr. Lawrence Rove, the late Mr. Peter Henele, Mr. Peter Job, the late Rev. Nathan Kera, Mr. David Kera, the Late Sir Ikan Rove, and the Hon. Job Dudley Tausinga. I am also grateful to the staff of the ministries of Forest, Environment, and Conservation, of Education and Human Resources Development, of Health and Medical Services, and the government of Solomon Islands, and especially Mr. Myknee Qusa, Mr. Basile Gua, Mr. Lawrence Fo'ana'ota, Ms. Freda Pitakaka, Mr. Lawrence Kiko, and Mr. Makiva Tuni.

My sincere thanks also go to my academic mentors, colleagues, and friends, especially Prof. Ryutaro Ohtsuka (Japan Wildlife Research Center), Prof. Shankar Aswani (Rhodes University), Prof. Yutaka Suga, Prof. Takeshi Matsui, and Prof. Masashi Haneda (all of The University of Tokyo), Prof. Shigeo Kobayashi, Prof. Akihisa Iwata, Prof. Shinya Takeda, and Dr. Hiroaki Setoguchi (all of Kyoto University), Dr. Daichi Ishimori (Musashi University), Dr. Motomu Tanaka (Kyushu University), Dr. Taishi Midorikawa (Obirin University), Dr. Takuya Nagaoka (Archaeological Institute of Kashihara), Prof. Chiho Watanabe and Dr. Masahiro Umezaki (both of The University of Tokyo), Dr. Jiang Hong Wei (Research Institute for Humanity and Nature), Dr. Taro Yamauchi (Hokkaido University), Dr. Minato Nakazawa (Kobe University), Dr. Simon Albert (Queensland University), Dr. Akitoshi Iwamoto (Tokyo Gakugei University), Prof. Kazuhiko Takeuchi (United Nations University), Dr. Kiyoshi Tadokoro (Akita University), and Mr. Masao Yamagata (Solomon Kitano Mendana Hotel). Mr. Philip E. Hyatt and other staff of Edanz Group, Ltd. proofread and edited the English of this book and greatly improved the text for publication.

I would like to thank Prof. Yoko Hayami and Dr. Narumi Shitara and other members of the Editorial Office and Editorial Committee of the Center of Southeast Asian Studies, Kyoto University and three anonymous referees for their guidance and useful comments.

The "Harmonization of Rural Development and the Community Welfare" (leader: R. Ohtsuka) program of the Japan Society for Promotion of Sciences and a KAKENHI Grant-in-Aid for Scientific Research (T. Furusawa) financially supported this study.

This book employed new analyses and debates. However, after substantial expansion and revisions the following previously published papers provided part of the data on which Chaps. 3, 4, 5, 6, 7, 8, 9, and 10 are based:

Furusawa T (2004) Minzoku Chishiki ni Motoduku Ningen-Shokubutsu-Dobutsu no Kankei [Human-plant-animal relationships analyzed as folk knowledge]. In: Ohtsuka R (ed) Solomon Shoto: Saigo no Nettairin [Solomon Islands: the last frontier of tropical rain forest]. University of Tokyo Press, Tokyo, pp 55–81. (In Japanese)

Furusawa T (2009) Changing ethnobotanical knowledge of the Roviana people, Solomon Islands: quantitative approaches of its correlation with modernization. Hum Ecol 37(2):147–159

Furusawa T (2011) Ecological and economic analyses of urbanization in the customary lands of the Solomon Islands. People and Cult Ocean 27:1–17

Furusawa T (2012) Tracking fishing activities of the Roviana population in the Solomon Islands using a portable global positioning system (GPS) unit and a heart rate monitor. Field Methods 24(2):216–229

Furusawa T, Ohtsuka R (2009) The role of barrier islands in subsistence of the inhabitants of Roviana Lagoon, Solomon Islands. Hum Ecol 37(5):629–642

Furusawa T, Sirikolo MQ, Sasaoka M, Ohtsuka R (2014) Interaction between forest biodiversity and people's use of forest resources in Roviana, Solomon Islands: implications for biocultural conservation under socioeconomic changes. J Ethnobiol Ethnomed 10:10

My final sincere thanks go to my family, Hana, Kenshun, and Gen.

Kyoto Takuro Furusawa

Conventions

Spelling and Wording

The spelling of the word "Roviana" in this book follows standardized orthography (Waterhouse 1928): the letters *b* and *d* are pre-nasalized and pronounced [mb] and [nd] as in English, a nasal velar stop is used to pronounce *ŋ* [ng] as in English, and a nasal velar stop, further pre-nasalized for *q*, is pronounced [ngg] as in English.

"Horticulture" is defined as the small-scale agriculture practiced by the people of the Solomon Islands, and "gardens" refer to agricultural fields, following conventional terminology in other research discussing Oceania.

Please note that Solomon Islands, as a country, does not use the word "The" as part of the country's name. However, the author does say "the Solomon Islands" when referring to the region and not the country.

Notes on Methods and Analyses

The author has lived in Roviana for more than 2 years since 2001, speaks the local Roviana language, understands the local customs and culture, and has built a rapport with the local people. The ministries of Education and Human Resources Development, of Environment, Conservation and Meteorology, and of Health and Medical Services of Solomon Islands permitted and supported this study.

Botanical specimens were collected in the field, and Myknee Qusa Sirikolo (current Director of the National Herbarium and Botanical Gardens of Solomon Islands, BSIP) at the Poitete Institute of Forestry, Western Province, Solomon Islands identified the specimens using their scientific names. Local Roviana names were used as species designations in analyses to connect the vegetation and plant use data, although, in some cases, the villagers recognized two or more taxonomic species as one Roviana vernacular name or vice versa. English common names were used for convenience. Therefore, the word "species" as used here refers to a unit of the local vernacular name rather than "species" as a scientific name, although in most cases both types of names refer to the same scientifically named species.

All research was performed after obtaining informed consent that included a requirement that all personal information would be analyzed anonymously and not made available to the public. Additionally, this book does not include some details of people's behavior and knowledge in an effort to protect their privacy as well as to protect intellectual property rights. I declare no conflict of interests.

Statistical significance was assumed when the P-value was less than 0.05.

Currency

1 Solomon Islands Dollar (SBD) $\approx$ 0.14 United States Dollars (USD) at the time of the study (2003 for the economic study). Note that the exchange rate dramatically changed between 2001 and 2003; since 2003 the exchange rate has stabilized and rarely moved from 2003 levels. Cash economic data used in this study were collected in 2003.

Contents

Abbreviations

AD	*Anno domini*
AN	Austronesian
BC	before Christ
BMI	body mass index
BSIP	National Herbarium and Botanical Gardens of Solomon Islands
CBD	Convention on Biological Diversity
CFC	Christian Fellowship Church
COP	Convention of the Parties
CU	consumer unit
DBH	diameter at breast height
FAO	Food and Agriculture Organization of the United Nations
GIS	geographical information system
GPS	global positioning system
IUCN	International Union for Conservation of Nature
MJ	mega joule
MPA	marine protected areas
NAN	non-Austronesian
PU	producer unit
REDD+	Reducing Emissions from Deforestation and forest Degradation in developing countries and the role of conservation, sustainable management of forests and the enhancement of forest carbon stocks in developing countries
SBD	Solomon Islands Dollars
SDA	Seventh-Day Adventist
USGS	United States Geological Survey

Chapter 1
Biocultural Diversity in Melanesia

Abstract The book starts with a review of background information related to East Melanesian Biodiversity Hotspot—especially the Solomon Islands, human adaptation and biocultural diversity, as well as socioeconomic changes. Many studies have documented the rich knowledge of the local peoples and their use of local floral or faunal and other aspects of the region's biological and physical environments. Recently, however, environmental anthropologists have debated whether traditional societies have voluntarily intended to preserve the natural world and whether their knowledge and practices have been adapted to the purpose of environmental preservation.

The goal of this book is to document the practices and knowledge of the Roviana people in relation to their use and management of biodiversity while using an interdisciplinary approach. In so doing, this book addresses the following general questions: "What kinds of factors have determined sustainability versus loss of biocultural diversity in Solomon Islands?" "How do the people of Solomon Islands recognize the conservation of nature and biodiversity in a traditional way or under socioeconomic change?" and "How can the society of Solomon Islands achieve such harmony under the currently changing socioeconomic conditions?" These questions are closely related with recent discussions taking place at the global scale.

Keywords Biocultural diversity • Conservation ethics • Ecological history • Socioeconomic changes • East Melanesian Biodiversity Hotspot • Structure of book • Study purpose

T. Furusawa, *Living with Biodiversity in an Island Ecosystem*, Ethnobiology,
DOI 10.1007/978-981-287-904-2_1

1.1 Diversity in Melanesia

1.1.1 Biological Diversity

Biological diversity, or simply biodiversity, refers to the variety of living organisms on Earth and is usually defined in terms of genetic, species, population, community, and ecosystem diversity (Groombridge and Jenkins 2002). Humans receive various services from the biodiversity present in the earth's plants and animals including the provisioning of food and medicine as well as aesthetic, economic, and several other types of benefits (Chivian and Bernstein 2008). However, human activities that lead to the degradation of land, water, and forests have contributed to a loss of such biodiversity. Recent discussion related to the loss of biodiversity has focused not only on the ecological consequences of this loss but also on the various benefits biodiversity provides to humans. In the context of developing societies, biodiversity plays an essential part in relieving human suffering and improving the quality of human life. Conservation of tropical rainforest is critical to preserving global biodiversity because these forests have high levels of natural diversity and are capable of storing large carbon stocks. Frequently, although such areas are largely uninhabited by humans, they have a high market value and generate economic interest from various organizations outside the local area.

The small islands of Oceania, including the Solomon Islands, are well known for their rich biodiversity. These islands characteristically have a high percentage of forest cover (Food and Agriculture Organization of the United Nations 2011) (Table 1.1). Five of the top ten countries with a high proportion of forest cover, the Federated States of Micronesia, American Samoa, Palau, Pitcairn, and Solomon Islands, are a part of Oceania; the Solomon Islands had the greatest percentage of forest cover when small island countries which had less than 10 km^2 of land area were excluded. Forests cover more than half of Papua New Guinea and Fiji. Table 1.2 shows the number of species known to occur in Australia and three Melanesian countries. Although these countries support a relatively small number of species, the proportion of endemic species is high. The ecosystem diversity of Oceania can be described as the diversity of this entire area, as well as the diversity of each island or each country.

Human impacts on biodiversity have increased in this region. Since humans first migrated to Oceania, they have used ecosystem services to provide themselves with subsistence through hunting, fishing, gathering, and horticulture. The human impacts on biodiversity in Oceania have dramatically increased with development during the past century, such as commercial logging. Furthermore, local people have increased their use of ecological services as the population grows and a market economy has developed, both causing increased use of natural resources. Deforestation rates in Papua New Guinea and the Solomon Islands are similar to or greater than the global average (Table 1.1). Additionally, a high proportion of species are now threatened as defined by the International Union for Conservation of Nature (IUCN) "Red List" (Table 1.2) (Groombridge and Jenkins 2002).

Table 1.1 Percent forest cover and rates of forest cover change in Melanesia and other countries worldwide

Rank	Country or area name	Land area (1000 ha)	% Forest area in 2010	Annual change rate (%)	
				1990–2000	2000–2010
1	French Guiana (South America)	8220	98	−0.1	0
2	Suriname (South America)	15,600	95	0	0
3	Federated States of Micronesia (Micronesia)	70	92	0	0
4	American Samoa (Polynesia)	20	89	−0.2	−0.2
5	Seychelles (East Africa)	45	88	0	0
6	Palau (Micronesia)	46	88	+0.4	+0.2
7	Gabon (Central Africa)	25,767	85	0	0
8	Pitcairn Island (Polynesia)	4	83	0	0
9	Turks and Caicos Islands (Caribbean)	43	80	0	0
10	Solomon Islands (Melanesia)	2799	79	−0.2	−0.2
	Papua New Guinea (Melanesia)	45,286	63	−0.4	−0.5
	Fiji (Melanesia)	1827	56	+0.3	+0.3
	World	13,009,550	31	−0.2	−0.1

Source: Food and Agriculture Organization of the United Nations (2011)

Conservation International (2012) registers the biogeographical region encompassing the East Melanesian Islands, including 1,600 islands, as one of the world's biodiversity hotspots (Fig. 1.1). These hotspots represent the richest and the most threatened reservoirs of plant and animal life on earth. These serve as the most important sites needing immediate attention for the conservation of biodiversity; a priority area for conservation activities means the area has a high percentage of endemic species and species at high risk of extinction. This region extends 99,384 km from the islands of Papua New Guinea, through Solomon Islands to Vanuatu. The region represents the most geographically complex area of islands on the earth and formed as a result of the varied geological age and stages of development of this landscape. This complex geological history allowed a unique flora and fauna to evolve within the entire hotspot and resulted in high levels of island endemism on individual islands. Large islands, such as New Britain and New Ireland, have mountainous terrain and were formed by geological uplift. Other medium-sized and small islands, such as Guadalcanal, Malaita, New Georgia, Santa Isabel, Makira, and the island of Choiseul Province in Solomon Islands, formed relatively recently

Table 1.2 Total number of mammal, bird, and plant species and the number of endemic and threatened species in these groups in representative countries of Oceania

Country	Mammals			Birds			Plants	
	Total	Endemic	Threatened[a]	Breeding total	Endemic	Threatened[a]	Total	Endemic
Australia	252	206	63	649	350	32	15,638	14074
Papua New Guinea	214	65	58	644	94	32	11,544	N.D.
Solomon Islands	53	21	21	163	43	23	3172	30
Fiji	4	1	4	74	24	12	1518	760
Worldwide	4629	–	1096	9672	–	1107	270,000	–

Source: Groombridge and Jenkins (2002)

[a]Defined by the International Union for Conservation of Nature (IUCN) "Red List"

Fig. 1.1 Location of the East Melanesian Islands Biodiversity Hotspot and Melanesian countries; PNG represents Papua New Guinea (Map source: Wikimedia Commons)

as volcanic islands. Habitats include coastal vegetation, mangrove forests, freshwater swamp forests, lowland rainforests, monsoon forests and grasslands, and montane rainforests. The islands of the region support approximately 8,000 plant species (of which 37.5 % or 3,000 are endemic), 86 mammal species (45.3 % or 39 endemic species), and 360 bird species (41.4 % or 149 endemic species (41.4 %)). The IUCN Red List of threatened animals includes 20 mammals and 33 birds of the above species.

1.1.2 Cultural Diversity

Melanesia is well known for its cultural and linguistic diversity. For example, the Summer Institute of Linguistics (SIL) has recorded 836 and 70 living languages in Papua New Guinea and Solomon Islands, respectively (Gordon Jr. 2005), most of which are unintelligible to each other. For example, in Solomon Islands, although the national language has been English since British colonization, Solomon pidgin (also spelled as Solomon Pijin, the lingua franca of the country) is recognized as the common language. Nevertheless, the people still generally speak their own local languages, and they only use pidgin for communication with other language speakers and rarely speak English. However, migration and exogamic marriage between different language populations have increased recently; this has slightly and gradually changed the language situation. Additionally, the increasing use of pidgin and English may have resulted in a decrease in the linguistic links between traditional culture and the natural world (Conservation International 2012).

A brief explanation of the roots of the local cultural diversity is necessary. Linguistic, archaeological, and genetic studies have previously suggested that the current human populations of Oceania are derived from an Austronesian (AN)-speaking people who migrated from continental Asia. The people of this migration, which started roughly in BC 4000–3000, passed through the islands of Southeast Asia and Melanesia, arriving eastward in Hawai'i and Easter Island around AD 1200. These people have now settled in Polynesia, Micronesia, and parts of Melanesia; however, the majority of the inhabitants of New Guinea and its neighboring islands in Melanesia are an exception. Specifically, they are non-Austronesian (NAN)-speaking and may be descendants of the first human population who settled in the region around BC 50,000–30,000 (Diamond 1988; Friedlaender et al. 2008; Oppenheimer and Richards 2001). Although the location of the original landing site of an AN-speaking people is under debate, as is the time of their arrival, linguistic and archaeological studies suggest they arrived in Taiwan from inland China roughly in BC 6000. Melanesia thus represents a unique region globally where groups of humans settled and expanded across such a wide geographical area over a long period of time. This region is also important for scientists studying human biological (genetic) adaptation, although this topic is beyond the scope of this book (e.g., Friedlaender et al. 2008; Furusawa et al. 2010; Furusawa et al. 2011b, 2013; Kimura et al. 2008; Lum et al. 2002).

Modes of subsistence differ from one society to another in this region. Foraging techniques, such as hunting, fishing, and gathering, have been commonly practiced throughout the region, although the intensity varies from place to place. For example, peoples in Highland or Lowland New Guinea Island have been known as hunters, and those on small coastal islands tend to be fishermen, although the former also practice small-scale fishing and the latter infrequently go hunting (Hviding and Bayliss-Smith 2000; Kawabe 1983; Morren 1986; Ohtsuka 1983). Additionally, Allaby (2007) classified three types of plant exploitation related to subsistence cultures while discussing this region as a center of plant domestication. The first type, "vegeculture," primarily exploits corms, rhizomes, and roots of species of taro, yam, sweet potato, and other root crops (Fig. 1.2). These crops with storage organs provide a suitable food source for people in tropical environments in the same way that cereals provide food for people in temperate regions. Taro and yam serve as the core of traditional diets in most islands in the Pacific (Fig. 1.2), except where sweet potato is produced through intensive agriculture in the highlands of Papua New Guinea (Bayliss-Smith 1974; Hviding 1996, 2005; Bayliss-Smith and Hviding

Fig. 1.2 A large tuber of a kind of taro, photographed at Olive in 2002

2012; Morren 1986; Rappaport 1984; Sillitoe 1996; Umezaki et al. 2000). People domesticated taro, an indigenous plant of New Guinea, in the early Holocene; taro was also domesticated in Southeast Asia. Two hypotheses exist related to the first introduction of sweet potato (*Ipomoea batatas*) into Melanesia. These hypotheses suggest sweet potato may have come from prehistoric imports from Latin American people or from historic Europeans. However, the cultivation of sweet potato in the highlands of New Guinea predates European contact. Clearly, this crop as well as cassava were recently introduced by Europeans (such as by governors or missionaries) to most other island societies such as Solomon Islands.

The second theory involves "arboriculture." Native people harvest nuts and fruits from a wide range of tree species. Trees such as *Canarium* nut (Fig. 1.3), coconut (*Cocos nucifera*), cutnut (*Barringtonia* spp.), sea almond (*Terminalia catappa*), screw pine (*Pandanus* spp.), sandal wood (*Santalum* spp.), and Fijian longan (*Pometia pinnata*) are important food sources and were domesticated in Melanesia. Common arboriculture trees in Polynesia such as breadfruit (*Artocarpus altilis*) and Tahitian chestnut (*Inocarpus fagifer*) had been relatively rare until recently (Pollock 1992). The first indication of plant exploitation in Melanesia dates back to BC 12,000 when people used *Canarium* nut (Yen 1990), and archaeological evidence

Fig. 1.3 *Canarium* nut tree growing in a settlement at Olive, in 2001

indicates NAN-people had cultivated many other plants before the arrival of AN-speaking people. Coconut is well adapted for oceanic dispersal, because its fruit is well protected and it can remain dormant for a long time. People domesticated this plant multiple times and in multiple places in the Pacific. AN-people also introduced various crops such as Malay apple (*Eugenia malaccensis*), Tahitian apple (*Spondias dulcis*), and probably betel nut (*Areca catechu*) from Southeast Asia.

"Herbiculture," cultivating non-root and non-tree crops such as bananas (*Musa* spp.) and sugarcane (*Saccharum* spp.), has been frequently classified as arboriculture. Archaeological studies (Allaby 2007) indicate people used banana in the early Holocene and cultivated it starting roughly in BC 5,000–4,000. In addition to these three categories of vegeculture, arboriculture, and herbiculture, the use of wild or semi-domesticated plants such as sago palm (*Metroxylon* spp.) also provides an important food source. Starch of sago trunk provides people with their main source of energy in lowland areas (Ohtsuka 1983).

In Melanesia and the Islands of Southeast Asia as well as in several other tropical countries, shifting cultivation societies have developed their own strategies for land use (Conklin 1972; Whitmore 1998). A fallow period allows the natural regeneration of soil nutrients during shifting cultivation. In the past, people cleared small patches of forest for cultivation and targeted secondary forest for use in shifting cultivation because this reduced the labor required to clear the land. Therefore, this mode of subsistence farming remained ecologically sound and sustainable as long as people only cleared small plots and used a short rotation cycle. Societies of this region, as with other societies in tropical rainforests, also depend on ecosystem services provided by forests for other forms of a subsistence lifestyle, including the collection of natural resource products, fishing, and hunting (Henderson and Hancock 1988; Hviding 1996; Hviding and Bayliss-Smith 2000). These peoples had a spiritual connection to the natural world and consequently learned to use forest resources sustainably (Hviding 2005; Kwa'ioloa and Burt 2001).

1.2 Socioeconomic Changes

In Melanesia, contact with Europeans caused rapid socioeconomic, demographic, and cultural changes (Aswani 1997; Bennett 1987; Hviding 2003b). Solomon Islands has recently experienced the early stages of economic development. In this country, the rapid increase in population and the introduction of a market economy occurred later than it did in many other developing countries. The growth in gross domestic product exceeded 4 % starting in 1990, although a period of ethnic tension (1998–2003) disrupted that growth rate. The population increased at an annual rate of 2.8 % during 1986–1999 and 2.3 % during 1999–2009. This exceeded the growth rates of most countries in the Asia-Pacific region (Solomon Islands National Statistics Office 2011), although the gross national income (Atlas method) per capita of 910 USD in 2009 for Solomon Islands was lower than the average of low-income countries globally (World Bank 2011). The rapid increase of the urban population of Solomon Islands (Table 1.3; Fig. 1.4) has exceeded the growth rate of other Asia-Pacific countries, and this growth rate has not diminished recently, unlike that of most other countries. However, urban households in Solomon Islands are not like urban households of large urban areas worldwide because many so-called urban people still need to depend on natural resources.

The expansion of agricultural lands, commercial logging, the development of timber and oil palm plantations as well as the construction of urban infrastructure reflect the recent increase in effects of human activities in Solomon Islands, while biodiversity has deteriorated in the region (Bennett 2000; Fazey et al. 2011; Hviding and Bayliss-Smith 2000). With the recent exposure of local people to Western culture, the demand for commodities has increased and continues to increase rapidly. Additionally, the export of natural resources has provided a main source of cash income in the region; therefore, each country has only limited motivation to place restraints on resource development (Bennett 1987; Myers et al. 2000; Solomon Islands Government 1999). While Solomon Islands expanded timber exports in the recent past, Malaysia and other Southeast Asian countries have passed new regulations resulting in a decrease in timber exports (Fig. 1.5). Environmental groups,

Table 1.3 Growth of the urban population in Solomon Islands in comparison with East and Southeast Asia and the Pacific, low-income countries worldwide, and the world[a, b]

	Solomon Islands	East/Southeast Asia & Pacific (developing countries only)	Low income	World
Estimated urban population (% of total)				
1979	22,777 (10.3)	281,809,131 (21.1)	76,449,353 (18.7)	1,683,864,738 (38.7)
1989	40,983 (13.4)	441,271,789 (28.0)	116,610,940 (21.9)	2,197,932,343 (42.5)
1999	62,684 (15.5)	647,622,272 (36.2)	168,908,902 (24.9)	2,769,114,174 (46.3)
2009	95,635 (18.3)	874,817,497 (45.0)	242,918,146 (28.7)	3,397,668,100 (50.3)
Urban population growth rate (%/year)				
1980–1989	5.8	4.6	4.1	2.7
1990–1999	4.2	3.8	3.7	2.3
2000–2009	4.2	2.9	3.6	2.0

[a]Source: (World Bank 2011)
[b]Urban areas were defined by the statistical office of each respective country

Fig. 1.4 A busy central market in Honiara, the capital of Solomon Islands, in 2009

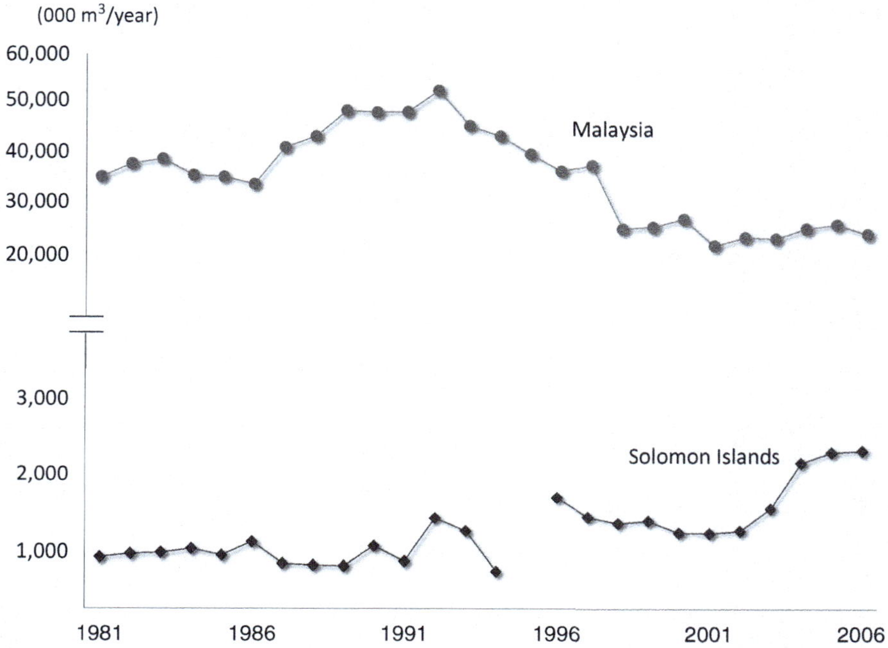

Fig. 1.5 Increase of timber exports from Solomon Islands, 1981–2006 (Data source: Asian Development Bank 2007a)

such as nongovernmental organizations (NGOs), had cooperated with government agencies to make some types of logging illegal (Roughan 1990), but legal logging continues to expand.

The country has been struggling to shift its industrial structure from a dependence on logging to the development of mining and agricultural industries (Allen and Sinclair 2013). However, Gold Ridge mine, which reopened in 2011 after the shutdown during a period of ethnic tensions, again slowed down its operation in 2014. Agricultural reforms have not met local needs, so that logging operations continue in many rural areas where logging formerly occurred a few decades ago.

Rapid population growth has intensified the traditional practice of shifting cultivation through the expansion of cultivated lands, a decreased fallow period, and an increased cultivation period (Umezaki et al. 2000). This shortening of the cultivation cycle generally results in degraded soil along with a loss of nutrients, decreased production per area, and deforestation (Bennett 2000; Lawrence et al. 1998; Whitmore 1998). Additionally, since the market economy has spread to societies that previously depended on a subsistence economy, these people have attempted to begin cash cropping; cash crops include coffee, rubber, oil palm, eucalyptus, and teak (Coelli and Fleming 2004; Hviding 1996; Hviding and Bayliss-Smith 2000; Lawrence et al. 1998; Ohtsuka et al. 1995). Such cash crops use perennial plants that occupy the land for long periods; therefore, the large areas of the land that are devoted to these perennial species become unavailable as fallow land or for cultivation in the traditional shifting cycles. Moreover, although traditional subsistence has generally taken place under communal management and use of land or resources, the market economy and population growth in tandem with the introduction of the European concept of land ownership have driven people to initiate private land ownership (Aswani 2002; Hviding and Bayliss-Smith 2000; Larmour 1997).

Recently, several academic and nongovernmental projects have worked to establish community-based conservation efforts that target natural resources. In Melanesian countries, most lands (e.g., 87 % of the land in Solomon Islands) are classified as "Customary Land" and are managed by traditional genealogical groups (Bennett 2000; Statistics Office 1995), so that rural communities have autonomy over their forests and manage them independently. Consensus among the groups that manage customary land and the integration of traditional ecological knowledge are recognized as important factors that should be considered if land conservation programs are to be successful (Aswani et al. 2007; Carter 1997; Read and Moseby 2006). In urbanized areas, however, land disputes have hindered the development of a consensus related to the protection of customary lands, allowing forest exploitation to continue.

Additionally, urban areas have been recognized as a haven for social unrest, as shown in the Honiara Riot (2006) and Ethnic Tensions (1998–2003); the high rate of unemployment that young people experience, disparities in capabilities and income, and degraded human relationships in towns are a few of the several reasons for this unrest (Liloqula 2000; Romer and Renzaho 2007).

1.3 Previous Studies in Melanesia

Ethnobiological studies—studies of the relationships between biological and cultural diversity—have been conducted in Melanesia. Many studies have documented the rich knowledge of the local peoples and their use of local floral or faunal and other aspects of the region's biological and physical environments (e.g., Akimichi 1978; Hviding 2005; Kwa'ioloa and Burt 2001; Pikacha 2008; Sillitoe 1998; Takekawa 2000; Nombo and Leach 2010; Hviding and Bayliss-Smith 2000).

Recently, however, environmental anthropologists have debated whether traditional societies have intended to preserve the natural world and whether their knowledge and practices have been adapted to the purpose of environmental preservation. Past anthropological studies in Melanesia have also generally declined to recognize a "traditional conservation ethic" in which local people are willing to preserve the natural world or local ecosystems (Bulmer 1982; Dwyer 1994), even though Western knowledge and technologies have obviously threatened to alter natural conditions recently (Colin Filer 2011). The local narratives stating that the local indigenous people had traditional conservation ethics have been recognized as having been influenced by the views of Western conservationists; that is, even the existence of such knowledge and practices has been rejected and the idea that local peoples had these characteristics has now been recognized as the "myth of the ecologically noble savage" and as a "populist" view (Redford 1991; van Helden 1998; Macintyre and Foale 2002; Filer 2004; West 2006; Wagner 2007; Colin Filer 2011). To some extent, this view, this idea that local indigenous people practiced conservation ecology, was an outgrowth of the desires and interests of some Western conservationists who were activists in protecting the environment (Dwyer 1994; Carrier 2010), and the same idea had clearly been used by the local people to claim their rights to resources, compensation, and royalties in relation to the claims of companies interested in logging or mining forest lands (Martha Macyntire and Foale 2002; Kabutaulaka 2006).

Smith and Wishnie (2000) reviewed and cited 196 previous reports from around the world related to this topic. They concluded that among traditional indigenous societies, even voluntary effective conservation was rare; however, resource management and techniques that result in the sustainable use of resources have been common and have indirectly resulted in the conservation and creation of biodiversity in traditional societies. Therefore, a lack of ethics or activities related to resource exploitation in these societies is not necessarily causing ecosystem destruction or degradation. One representative case is that of Bayliss-Smith and others (2003) who reported a situation where the biodiversity of a forest was a product of large-scale artificial modification of that forest by local peoples in the past on New Georgia Island, Solomon Islands. Overall, to successfully and sustainably conserve biodiversity, one must first understand local knowledge and practices and their direct and indirect effects on biodiversity; then, these can be integrated with contemporary scientific knowledge into creating a system of management that employs community-based and community-participatory methods (Cohen et al. 2014).

1.4 Living with Biodiversity

The goal of this book is to document the practices and knowledge of the Roviana people in relation to their use and management of biodiversity while using an interdisciplinary approach. In so doing, this book tries to address the following general questions: "What kinds of factors have determined sustainability versus loss of biocultural diversity in Solomon Islands?" "How do the people of Solomon Islands recognize the conservation of nature and biodiversity in a traditional way or under socioeconomic change?" and "How can the society of Solomon Islands achieve such harmony under the currently changing socioeconomic conditions?" These questions are closely related with recent discussions taking place at the global scale. The Convention on Biological Diversity (CBD) has highlighted the importance of integrating biodiversity conservation with efforts to protect the rights of indigenous peoples to use their ecosystems to improve their lifestyles (Adams et al. 2004; Myers et al. 2000; Sanderson and Redford 2004; van Oudenhoven et al. 2010). In contrast, conservation decisions by agents outside of those communities have often ignored local needs, causing disputes with local people and achieving limited success as described in the previous section.

Proponents of biodiversity conservation have emphasized that related efforts should aim to preserve sustainable human-modified natural environments, also called social–ecological production landscapes (SEPL), by encouraging broader global recognition of their value (e.g., Target 3 in the Strategic Plan for Biodiversity 2011–2020 and the SATOYAMA Initiative in the 10th Meeting of the Convention of the Parties, Convention on Biological Diversity (CBD-COP10), 2010) (Bélair et al. 2010; Takeuchi 2010). These traditional societies have often contributed to, rather than hindered, the creation and conservation of biodiversity. Frequent human removal of trees aided the growth of pioneer tree species, providing an ecological niche for some kinds of animals and contributing to forest regeneration. Human activities, such as agriculture, forestry, fishing, and hunting helped to form a diversified landscape, which cultivated the existence of various floral and faunal species in that landscape while simultaneously enhancing ecological services for the human forest inhabitants themselves by providing foods, fuel, and other materials. Recently, however, humans have converted many forests into more uniform, efficient, and large-scale production systems that often result in environmental degradation as well as the loss of both ecological services and local traditions. In contrast, the positive human impacts in social–ecological production landscapes are difficult to study methodologically and have often been neglected or undervalued in conservation programs. In this context, to avoid the overexploitation of forest resources, ecological services must satisfy the society's needs, and society must recognize this to achieve consensus on their protection and conservation. Therefore, addressing the extent to which people traditionally used ecosystem services and defining any such loss as it relates to overexploitation is vitally important.

Each chapter of this book discusses different types of biological or cultural diversity: i.e., Chap. 3 addresses landscape diversity, Chap. 4 discusses subsistence diversity, Chap. 5 resource use diversity and diversity of human–forest interactions, Chap. 6 diversity in environmental understanding, Chap. 6 knowledge diversity, Chap. 8 socioeconomic diversity, and Chap. 9 diversity in conservation activities. Through them, this book emphasizes the importance of understanding internal diversity; internal diversity includes the types of diversity which existed within a society or within an individual at various domains of human life, recognition, and knowledge associated with surrounding natural environment. The chapter structure of this book shows how people hold very diverse cultural and psychological attitudes toward biodiversity. A final discussion on the possible future of human–natural world interactions is thus made in terms of internal diversity and the roles of external agents.

For these purposes, this book adopted multiple levels of analyses according to the target: population-level analysis for food production and consumption efficiency, landscape management, and resource use within an ecosystem, as well as household-level analysis of production and consumption strategies and socioeconomic inequality, and individual-level analysis of psychological factors. This book is thus a unique study integrating interdisciplinary methods and concepts and multiple analyses of human ecology.

Chapter 2
Ecological History and Regional Context of Roviana

Abstract This chapter summarizes the geographical, ecological, social, and cultural background of the study site of the Solomon Islands. The high level of biodiversity found to persist there was perceived to have resulted from a lack of human intervention, but archaeological and forest ecological studies have indicated that many of the very old forests that make up the ecosystem in this area had once been cleared by ancestral people. Furthermore, the geological and vegetative characteristics of each island vary widely; this is especially true of the large volcanic island and small barrier islands, even within the limited geographical scale. People depend on root crops as their main source of food. Subsistence gardens, abandoned second-growth forests, and coconut plantations have rendered the landscape a mosaic of various land cover types. All Roviana people share the same culture and similar ecological settings, although the degree of modernization varies among them. Therefore, this area provided an ideal setting for this study's framework.

Keywords Geography • Society • Culture • Ecosystem • History • Study site

2.1 Fieldwork in Solomon Islands

Solomon Islands consists of a double chain of islands that include more than 900 small islands (Fig. 2.1). The country is a high-priority area for biodiversity conservation because of its location in the larger biodiversity hotspot of the East Melanesian Islands (Brooks et al. 2006; Conservation International 2012). Initially, the high level of biodiversity found to persist there was perceived to have resulted from a lack of human intervention, but archaeological and forest ecological studies have indicated that many of the very old forests that make up the ecosystem in this area had once been cleared by ancestral people in the Western Solomon Islands (Bayliss-Smith et al. 2003). Furthermore, the geological and vegetative characteristics of each island vary widely; this is especially true of the large volcanic island and small barrier islands, even within the limited geographical scale of Solomon Islands itself (Wall and Hansell 1975). The land use patterns also differ from one island to another (Aswani and Vaccaro 2008; Furusawa and Ohtsuka 2009; Hviding and Bayliss-Smith 2000).

T. Furusawa, *Living with Biodiversity in an Island Ecosystem*, Ethnobiology,
DOI 10.1007/978-981-287-904-2_2

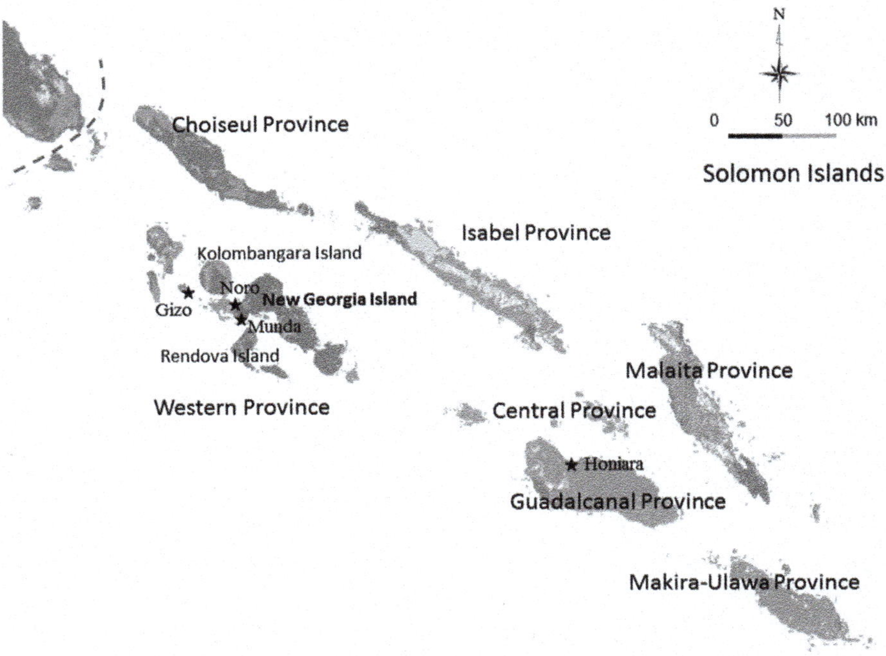

Fig. 2.1 Map of Solomon Islands and the location of Western Province

Traditional land tenure ownership is in effect on about 87 % of the lands. Members of a land-owning clan typically use this type of "customary land" exclusively (Bennett 2000). In these lands, the people must cope with socioeconomic and ecological changes without expecting any support from the government; unlike most other countries, but similar to other Melanesian countries, the government has very limited influence on urban planning. The people are allowed to use lands including barrier islands as long as they belong to the clan that traditionally manages that land. However, in and around urbanized areas, i.e., the center of demographic and economic changes, each household often establishes its own boundaries within the customary land and uses the land as their own *de facto* private territory, instead of resorting to communal use of the clan's land (Aswani 1999; Schneider 1998). This tendency has begun to spread to the rural areas.

The fieldwork was conducted in the Western Province of Solomon Islands from January to October 2001, from May to July 2002, from July 2003 to February 2004, in September 2004, from July to September 2005, in November 2006, in May and July 2007, in February 2008, in February and August 2009, in February 2010, and in November 2011. All these research efforts were conducted in the Roviana language with the help of two local assistants, Mr. Edwin Huti Vudere and Mr. Rex Dagha.

2.2 Geography and Ecosystem

Roviana lies in the southwestern part of New Georgia Island, near barrier and small islands, and extends for 150 km from Konggu Kalena Bay to Parara (Vonavona) Island in Western Province of Solomon Islands (Fig. 2.2). Geographically, New Georgia Island is covered by tropical forest that remains constantly mesic-wet, with average maximum and minimum temperatures of 30.3 °C and 24.2 °C (in 1993), respectively. The mean annual rainfall is 3458 mm, with little variation in monthly rainfall, ranging from 225 mm in October to 390 mm in February at Munda (Fig. 2.3), about 30 km east of Olive village (data from 1987 to 1993, Solomon Islands National Statistics Office 1995).

Extensive tropical lowland forests and swamp forests or mangroves originally covered New Georgia Island and barrier islands in inland areas and on the coast, respectively (Fig. 2.4). Subsistence gardens, abandoned second-growth forests, and coconut plantations have rendered the landscape a mosaic of various land cover types.

Commercial logging operations by overseas companies began in the western part (Vonavona) of Roviana in the 1960s, although most clans in the eastern part did not allow this type of logging until the 1980s (Fig. 2.5). The impact of logging was reflected not only in the forest environment but also through the change in lifestyles of the local peoples, because logging operations provided the villagers with employment, royalties, and materials for infrastructure, such as schools and medical aid stations (called "aidposts" in the country). Therefore, the difference in time of the onset of logging operations enhanced the variation in modernity in this area.

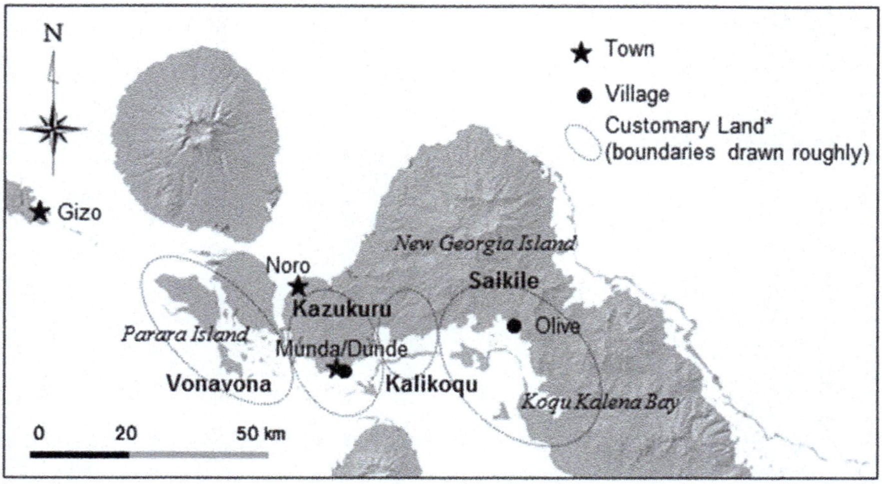

Fig. 2.2 Map of Roviana and Vonavona Lagoons (Data source: (USGS 2004). *There may be two or more sub-customary land units in each customary land unit; these are grouped into larger groups for convenience)

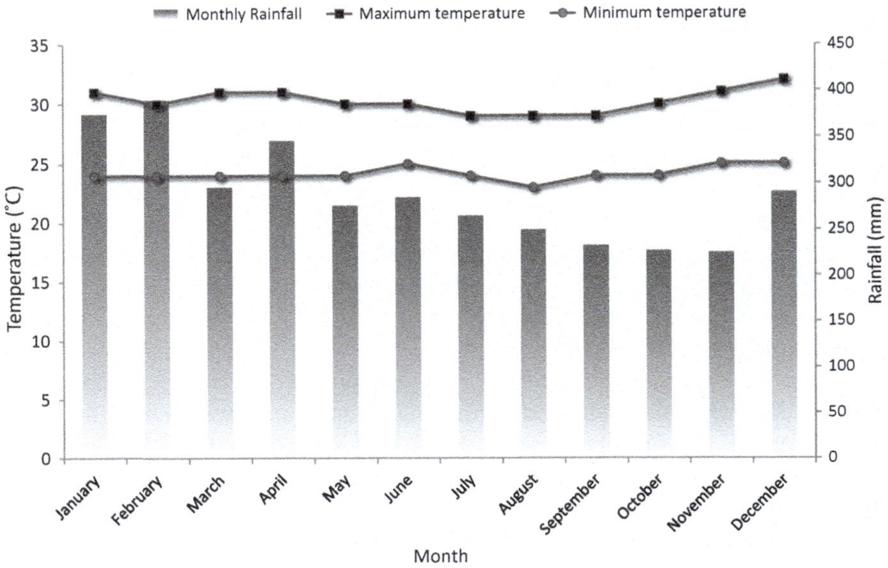

Fig. 2.3 Temperature and rainfall at Munda Meteorological Office (Munda Airstrip), New Georgia Island, based on data from 1987 to 1993 (Solomon Islands National Statistics Office 1995)

Fig. 2.4 A gigantic pandanus plant in swamp forest at Koqu Kalena Bay in 2003

Fig. 2.5 Round logs harvested from forest at Koqu Kalena Bay in 2009

Reforestation with imported commercial timber species such as eucalyptus (*Eucalyptus deglupta*) and teak (*Tectona grandis*) involves some of the most intense activities conducted for cash earnings rather than for ecological purposes (Fig. 2.6). A few decades will have to pass before the villagers will be able to harvest and sell the planted timber species. In 2001, a foreign research team designated marine-protected areas (MPAs) in the territories of Roviana villages, and villagers have been restricted from acquiring resources in these MPAs since 2002. These MPAs were designated in ecologically important areas in an attempt to minimize the effects of humans on fish species; in reality, no adverse effects, such as a decrease of fish catch or an increase in the physical labor required of fishermen to harvest fish, have ever been observed.

2.3 People of Roviana

The people of Roviana speak the Roviana language, one of the four languages spoken in the island of New Georgia (Fig. 2.2). Although the Roviana language and three other languages—Kusaghe, Marovo, and Kalikolo—belong to the West New Georgian language group and exhibit some overlap in vocabulary and grammar, they are mostly mutually unintelligible, and the vernacular names of plants and

Fig. 2.6 Timber plantation of *Eucalyptus deglupta* at Dago in 2007, 6 years after planting seedlings

animals differ from one language to another. Oral histories suggest that all Roviana clans descended from one ancestral group (Aswani 2000; Aswani and Sheppard 2003; Sheppard et al. 2000). Archaeological evidence and oral histories indicated increasing political control by the chiefs of Nusa Roviana, a small island located in central Roviana geographically, during the nineteenth century.

The Roviana people, 14,805 in number in 2010 (Solomon Islands National Statistics Office 2011), are semi-subsistent (Fig. 2.7). The most important form of subsistence agriculture in the Roviana region is the shifting cultivation of root crops, such as sweet potato (*Ipomoea batatas*), cassava (*Manihot esculenta*), taro (*Colocasia esculenta*), and yams (*Dioscorea* spp.), with a rotation of lands between cultivation and fallow ground. While they do go fishing, the Roviana people also spend a significant amount of time earning cash through crop farming and the collection of marine resources. Before the adoption of sweet potato and cassava as staples in the early twentieth century, this cultivation system using cropland rotation required an extensive area of land for cultivation and fallowing; fallow land exceeded the amount of land currently under cultivation by several times to allow the regeneration of trees and refertilization of the soil (Furusawa and Ohtsuka 2009; Whitmore 1998). Forests have provided various kinds of resources, such as building materials, medicinal plants and tools, and products used for magical/ritual purposes.

Westernization or modernization began in the nineteenth century when European explorers first contacted the people of Roviana. The labor trade, called blackbirding, in which local people were recruited or even kidnapped as workers for plantations in Fiji or Australia owned by Europeans, had affected the area since the 1860s (Campbell

Fig. 2.7 A settlement facing a calm lagoon at Nusa Banga in 2005

1989). The introduction of iron tomahawks, rifles, and other weapons was said to have enhanced "head-hunting" warfare. Additionally, Europeans brought infectious diseases to this area that were often fatal to local people who had no natural immunity to fight those diseases (Bayliss-Smith 2006). Accordingly, Roviana societies experienced rapid social changes and depopulation in the latter nineteenth century.

The Westernization further intensified after the attack on Nusa Roviana by the HMS Royalist in 1891. The region was proclaimed a British protectorate in 1893 and Christians established a mission in Munda in 1902 (Table 2.1). Munda has developed into a township since that time. An airstrip built during World War II also contributed to the commercialization of the area, resulting in the present population of 3000. Solomon Islands gained independence from Britain in 1978. Solomon Islands constructed an international port and a large fisheries base (Solomon Taiyo Limited) at Noro, 10 km north of Munda. Munda is the fourth largest town in Solomon Islands, following Honiara, Gizo, and Auki, the national capital, and the capitals of Western and Malaita Provinces, respectively. Munda lies 50 km east of Gizo. In this study, the three commercial centers of Munda, Gizo, and Noro are referred to as towns. Thus, the extent of modernization has varied within Roviana. The creation of a cash economy and the availability of Western materials have generally affected the villages located in Munda and to the west, i.e., villages near Munda, Noro, or Gizo. Most people in Roviana live in Western-style permanent houses. In contrast, the development of towns has had less influence on villages in the east, i.e., the more remote inner lagoon areas that lie far from the towns. These

Table 2.1 Historical chronology of Roviana and the Solomon Islands

Year	Event
~1600s	Establishment of Nusa Roviana chiefdom (Sheppard et al. 2000)
1860s	Onset of labor trade (blackbirding)
1880–1990s	Rapid depopulation from introduced diseases
1891	Attack on Nusa Roviana by HMS Royalist
1893	Proclamation of British Solomon Islands
1902	Methodist missionaries arrive and establish a mission
1941	Start of WWII
1942–1943	Occupation of New Georgia Island by the Japanese Army
1945	End of WWII
Around 1960	Christian Fellowship Church (CFC) separated from Methodist Church
~1960s	Commercial logging in Vonavona
1978	Independence of Solomon Islands
~1980s	Commercial logging begins in Eastern Roviana
1997–2003	Ethnic tension on Guadalcanal Island

people generally live in houses made from sago palm leaves. In contrast, even in the present day, people outside of Munda have few opportunities to obtain a cash income because people in the villages in Roviana Lagoon can only access Munda by boat.

All Roviana people share the same culture and similar ecological settings, although the degree of modernization varies among them. Therefore, this area provided an ideal setting for this study's framework.

People depend on root crops as their main source of food. Traditionally, people consumed various species of root crops including taro (Araceae) and yams. In the twentieth century, Christian missions introduced sweet potato and cassava which have become the main crops. People were introduced to the concept of a cash economy during the nineteenth century through trade of copra, turtle shells, trochus, and ivory nuts; later, the influence of cash economy expanded through the development of coconut plantations and copra production until the 1980s (Hviding and Bayliss-Smith 2000). Since the 1980s, royalty payments from logging and wage labor provided by logging camps became the main sources of cash income, and the economic impact of logging along with the increased importation of stable foods allowed people to depend on imported rice and flours. Additionally, they enjoyed the taste of rice. The shorter cooking time (<1 h) and smaller amount of firewood required to prepare rice makes it a more convenient food than root crops. The villagers commonly purchased biscuits and increasingly consume doughnuts and ban, two food products made from imported flour.

Villagers now cultivate slippery cabbage (*Hibiscus manihot*), an indigenous green vegetable, and other introduced food made from various species. Fish serve as a main source of protein. People in rural villages catch fish and often trade some fish on the open market in towns. The rural villagers purchase tinned tuna fish (called *taiyo* in pidgin) more commonly than other foods, providing themselves with an inexpensive and stable source of nutrients (Barclay and Wakabayashi 2000).

Fig. 2.8 Stone-oven-baked foods with other fresh foods offered at a wedding feast at Tobo in 2003

Currently, they very infrequently hunt land animals, with wild pigs serving as the main target. Some villagers have domesticated chicken, dugs, pigs, and goats, but animal husbandry is only at small scale. Both wild and cultivated coconut (*Cocos nucifera*) provide people with an important source of nutrients. Mature coconut husk and young coconut juice are high caloric food items. For example, during a 1-day fishing trip, some villagers will eat only the fresh coconuts they collect from the islands they visit.

Motu, stone oven baking, provides a traditional method of heating food, although boiling food in a purchased steel pot has mostly replaced traditional methods. However, people still frequently use *motu* to prepare foods for feasts (Fig. 2.8). They wrap foods in banana leaves, or in leaves of other species of Musaceae or Zingiberaceae, and place them in a pit among heated stones. These foods, including sweet potato, cassava, pudding made from cassava, fish, pig meat, and other local items, are covered with heated stones followed by a covering of soil or banana leaves.

2.4 Social and Cultural Background

Current ecological and botanical research and the associated quantitative data serve as the basis for this book. However, the study site exhibits several unique cultural contexts. The following description and characteristics provide background information.

2.4.1 Kastom

Kastom (spelled *kasitomu* in the Roviana language) is a pidgin word, literally meaning "custom." This concept usually refers to local traditions, traditional culture, and customs, in contrast to the customs, concepts, ideas, and materials of Western societies. *Kastom* is sometimes only used to indicate traditional ritual or magical power. Scholars have argued that people recognized the word *kastom* as defining them as unique when compared with other communities and as allowing them to establish their own cultural identities (Hviding and Bayliss-Smith 2000; Ishimori 2001).

2.4.2 Chief

In the context of the Solomon Islands, the word "chief" refers to a traditional leader. Typically, the concept of chief indicates the generational transfer of its authority through genealogical relationships, and a chief serves as a leader of people using group-held customary land. This type of chief is called *baŋara* in the Roviana language. In communities with large areas of customary land or when clan groups are involved, multiple chiefs are under the control of a leading chief who is called a "paramount chief (*ŋati baŋara*)" (Fig. 2.9). Chiefs usually play important roles in land tenure and resource management; they may be the largest landowner or leader of the largest clan in an area of customary land; as a result, in their clan, they serve as the most important stakeholder for discussions with other clans in dealing with the logging issue. In another case with more recent context, when the church leadership recognizes a church leader or other person, then the villagers also call that person a chief. A village organizer, officially appointed as a person connecting villagers to the provincial or national government, is different from a chief, although these two designations may overlap. Even if a leader is not called a village organizer, recently those who are not direct descendants of the chief but who are respected for their Christian church activities or for their success in economic activities have played the roles of a chief in some areas. In such areas, the power of the former type of traditional chiefs is decreasing (Sekine 2001).

2.4.3 Wantok

The people of Papua New Guinea and the Solomon Islands still use *wantok*, a pidgin word meaning literally "one talk or one language" and conveying common linguistic or kinship bonds, to refer to someone with whom a person has a relationship of reciprocal obligation, as opposed to a stranger (Barclay 2004). Because the people speak about 70 languages in the country, strong ties link members within each language group as opposed to neighboring groups and give them a group identity.

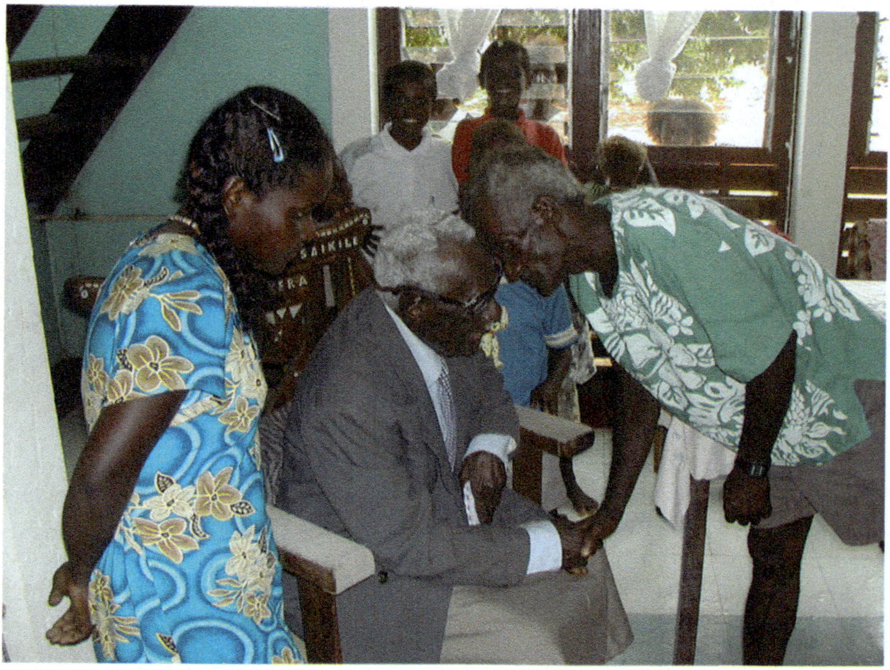

Fig. 2.9 The late Reverend Nathan Kera (sitting), the 8th Paramount Chief of the Saikile, is greeted by an elder at Vogi in 2003

Members in a language group usually develop mutually beneficial relationships and thus are likely to form a group in their original location as well as in urban areas. In many cases, these people genealogically or spiritually share the same ancestors; e.g., Roviana people believe they share a common ancestor.

2.4.4 Land Tenure

Land tenure is a very sensitive issue in the Solomon Islands. Land is owned by the clan (called *butubutu* in the Roviana language), unlike European systems of individual ownership, and this customary land is used exclusively by the members of the land-owning clan. However, definition of "clan" is sometimes dynamic because this depends on rules following patrilineal or matrilineal patterns as well as the time when the clan started. Additionally, land boundaries for different areas of customary land have only recently been drawn, so that the actual borders remain unclear. Further complicating the issue, ownership, and rights to use or manage land may vary from place to place. For example, the people consider some types of forest resources to be communal property regardless of the actual land ownership (Miyauchi 2001). Even if a person is not an owner or related to the ownership clan, that person may be allowed to live in a settlement or even make a garden on the land. Land tenure functions to provide exclusive use rights mainly in the case of commercial logging; someone harvesting timber may pay for the right to harvest timber based on land "ownership." Disputes frequently occur related to land tenure and the right to timber harvest. Note that clan ownership serves as the basis for sea tenure, although open access to marine resources is more likely because of the difficulty in controlling those resources; sea tenure is also transforming based on increasing market economy pressure (Akimichi 1991; Aswani 1999; Foale and Manele 2004). Follow-ups of land dispute cases have suggested that in many cases disputes are irresolvable (Foale 2001).

2.4.5 Tabu ples/Hope

Tabu ples (a pidgin word, literally meaning taboo place), called *hope* in the Roviana language, is a sacred place. Figure 2.10 illustrates a representative place where skulls of ancestors are enshrined (skull shrine). Human-made stones established in antiquity make such a *tabu ples* easily recognizable. Therefore, a *tabu ples* often provides evidence of land ownership in land disputes. Such sites are frequently used as tourism resources. Tourism may provide a potential industry in Roviana, but currently few tourists visit the area.

Fig. 2.10 A skull shrine of the Saikile people, located at Saikile Point, in 2001

One type of sacred place is called *hope maŋini* (Fig. 2.11). According to Nagaoka (1999, 2000), *hope maŋini* means “garden shrines”, which were one of the earliest and most fundamental shrines in addition to skull shrines. The priest (*hiama*) often prayed for the fertility of garden plots (*inuma*). A ritual called *sosope la pa reqe* was conducted before cultivating virgin forest and soil; ashes were ritually taken from the shrine to a proposed garden area (Nagaoka 1999, 2000). Chapter 3 also explains the concept of *hope*.

2.4.6 Tension

Tension describes, in a narrow sense, an ethnic conflict fought between indigenous Guadalcanal people and Malaitan immigrants located in the national capital of Honiara, and nearby areas on Guadalcanal Island, that started in 1998 and intensified in 2000. In 2003, a ceasefire was declared with the help of an international force called the Regional Assistant Mission to Solomon Islands (RAMSI), initially lead by units of the Australian army. In a broad sense, various conflicts between other

Fig. 2.11 A garden shrine (*hope maŋini*) at Dago in 2002

groups (*wantoks*) and relevant criminal issues caused by a lack of law-and-order during this period are also referred to as *tension* (Braithwaite et al. 2010).

2.4.7 Christianity

About 96 % of the people inhabiting Solomon Islands are Christians, and almost all people in Western Province belong to a Christian denomination (Solomon Islands National Statistics Office 2011). The first Christian missionary to visit this area, a Methodist, arrived in 1902 followed by members of the Seventh-Day Adventist (SDA) Church. A number of local population groups, who previously lived in both coastal areas and in inner mountainous areas, resettled in coastal villages where church priests were stationed. Within a few decades, but before World War II, all people were said to have converted to Christianity. Although the Methodists had been the largest sect, a local reverend who had been trained in Methodism founded a new sect called the Christian Fellowship Church (CFC) around 1960. This church partly integrated local beliefs and customs into the new church's theory and activities (Ishimori 2007; Nakazawa et al. 2002); the teachings of this group are said to include avoiding excessive participation in the cash economy and encouragement of traditional resource management. However, this sect also took the lead in several recent rural development projects (Chaps. 3, 9, and 10). After the CFC separated from the Methodist church, the majority of the Methodist followers joined the United Church. Accordingly, the United Church (38.7 %), SDA church (27.7 %), and CFC (14.2 %) are the largest church groups in the province (Rural Development Division 2001); the CFC is the largest in rural Roviana villages and North New Georgia areas, while other churches are dominant in towns, the Marovo area, and other islands. Note that the Church of Melanesia (Anglican: 31.9 %) has the most followers at the national level, followed by the Roman Catholic church (19.6 %), South Sea Evangelical Church (17.1 %), SDA church (11.7 %), United Church (10.1 %), and CFC (2.5 %) (Solomon Islands National Statistics Office 2011). Almost all people in Olive (98.4 %) were followers of the CFC; exceptional cases were those who immigrated to Olive after marriage. The people of Dunde supported various churches: United Church, 58.7 %; Methodist, 10.2 %; SDA, 8.7 %; CFC, 8.2 %; Jehovah's Witness, 3.7 %; Church of Melanesia (Anglican), 2.0 %; and others.

2.5 Main Villages Studied Here

A main target of this book was Olive village, consisting of 357 individuals living in 65 households (as of 2003) (Fig. 2.12). Olive, one of the largest villages in the Saikile Customary Land, lies approximately 32 km east of Munda town and lacks roads or public transportation. The Olive villagers who lived under the control of the

Fig. 2.12 Satellite (Landsat ETM+) image of Roviana and the approximate locations of the main villages studied here (NASA Landsat 7 Program 2008)

chieftaincy of the Saikile clan had rights to use some parts of the main island, New Georgia, and several barrier islands. The Saikile clan, one of the largest land-owning clans, settled in the eastern part of Roviana and had a population of 2068 (Statistics Office 1997); other small villages also lie scattered across the territory such as Ha'apai, Vogi, Nusa Hope, Baraulu, and others. Note that the Saikile Customary Land constituted the territory of the Saikile clan, but numerous land disputes have caused ownership of the land to be subdivided into smaller areas based on sub-clans. All inhabitants of the village were engaged in the shifting cultivation of root crops and fishing for their own consumption. The main source of cash income was selling marine resources such as *Nassarius* shells and sea cucumbers for export. In the 1980s, commercial logging started in this customary land, triggering changes by providing the people with employment and income from the sale of logging rights. The profits from the logging company, i.e., employment and royalty payments, have been decreasing since 2001 because of the removal of the logging campsite and intra-tribal disputes related to the logging operation.

A comparative study was conducted in Dunde settlement in Munda Township. Munda consists of a small government station area (Lambete) and five settlements (Dunde, Kekehe, Lodu Maho, Kokenggolo, Kindu) in the customary lands, which clans own and control; Dunde, the largest settlement (1065 inhabitants in 202 households in 2003), is divided into smaller segments. The urbanization of Munda town has long influenced the inhabitants of Dunde; the inner Munda area was the original home of the ancestors of all Roviana people and had been the political center before European contact. After the arrival of Europeans, this area became a center for trade between local people and Europeans in the nineteenth century. Christians founded their first mission there in the early twentieth century. During World War II, the Japanese army constructed a base and airfield, and after the war, the government developed the Lambete area for an airstrip, governmental offices, banks, and other commercial activities. Today, electricity, water, a public phone network, and several stores are available in Munda, not only on the government land but also in Dunde and other settlements. Land cover is changing from forest to cleared land, and urbanization has led to the pouring of concrete for some roads and in some settlements. These conditions satisfy the conventional definition of an urban area (McIntyre et al. 2008). However, the "urban" Solomon Islanders are still tied to their land base and often exhibit a degree of subsistence consumption; in reality, they could be better classified as only semi-urban (Furusawa 2011).

Modern life has affected Olive much less than Dunde. For example, in Dunde and Olive, 86 % and 28 % of the houses are built in a modern style, 93 % and 17 % of the households have rainwater tanks, 33 % and 11 % of households have outboard motors, and the average monthly household income is 1752 SBD (Solomon Islands Dollars; ≈ 245.3 USD) and 378 SBD (≈52.9 USD), respectively (Table 2.2). Only Dunde has both electricity and water lines. Saikile Customary Land is sparsely populated, with about eight persons per km^2. Estimating the population density for Dunde proved difficult, because several land disputes make a large portion of the customary land into de facto unavailable land. The author estimates the population density stands at more than 100 persons per km^2 based on observations of lands used by the people in this area. Meanwhile, the Asian Development Bank estimated 161 persons per km^2, which was the highest population density in the region (Asian Development Bank 2007b). Regardless of the large economic differences among the communities, oral history and archaeological evidence suggest that the people of Dunde and Olive are descended from a single ancestral population that migrated from Nusa Roviana Island, a small island located near Munda, in the late eighteenth or nineteenth century (Aswani 1999; Gordon Jr 2005). Additionally, ecological conditions are similar throughout the Roviana region.

In the Saikile Customary Land, areas near Olive village were logged in 1993–1994. While the logging negatively affected the forest, the local people profited from employment, royalty payments, and improved infrastructure. The forests surrounding the Dunde area were not logged because the rights to those forests were violently disputed within and among communities and clans. However, the people have benefitted economically from logging operations in nearby areas (e.g., Vonavona, Enoghae) since the 1960s through employment and royalty payments.

Table 2.2 Sociodemographic and modern features of the studied villages

	Dunde (Urban)	Olive (Rural)
No. of households (HHs)	202	65
No. of individuals	1065	379
Permanent houses per HH	0.86	0.28
No. of outboard motors per HH	0.26	0.11
No. of rainwater tanks per HH	0.86	0.17
No. of stores per HH	0.07	0.03

Chapter 3
Diversity in Landscape

Abstract This chapter describes the vegetation and landscape from both ecological and cultural viewpoints. Local people call the geographic setting of New Georgia Island *tutupeka* and that of the barrier islands *toba* and understand that the soil and vegetation differ on the barrier islands when compared with those of the main island. In the Roviana language, forest is generally called *hinqohinqo*, but 12 names were used to describe various forest vegetation types. Among them, *rizevu* (borrowed from the English word "reserve") is a forest reserve growing near the settlement. Logging by multinational companies and agricultural cultivation are prohibited in these areas, but the local people are allowed to harvest trees for personal use and to remove several non-timber forest products. Vegetation survey disclosed that the primary forests (*muqe*) on the main and barrier islands were only half as similar (16.0 %) as the respective secondary forests (*nobo*) (33.3 %). However, vegetation in the reserve forests was 45.9–52.6 % similar to that found in the primary and secondary forests on the main island. Thus, each forest type contained a different type of biodiversity, and this level of diversity is related to human activities.

Keywords Landscape classification • Vegetation survey • Primary forest • Secondary forest • Forest reserve • Logged forest • Mangrove • Similarity • Dissimilarity

3.1 Landscape Classification

To better understand the biological and cultural diversity of the society studied here, readers should first understand the landscape of the study site. Various previous studies have usually described landscapes from purely ecological or botanical viewpoints. However, such scientific classification or description of a landscape does not necessarily include a description of knowledge or behavior of local people. For example, a traditional classification of a landscape serves our purposes better than a simple scientific one because this helps the researcher to understand where people collect resources based on their knowledge and recognition. Therefore, bringing traditional knowledge into a botanical study will make it easy to match the

T. Furusawa, *Living with Biodiversity in an Island Ecosystem*, Ethnobiology,
DOI 10.1007/978-981-287-904-2_3

landscape being studied with the knowledge and behavior of local people. This integration also serves an important purpose in understanding how a society uses space for subsistence and life and provides a central concept in human ecology (Aswani 1997; Ellen 1982; Suda 1994). This chapter explains how the Roviana people classify landscapes and describes the diversity of their living environment. This includes an understanding of plant diversity within each vegetation type and diversity throughout the different vegetation types as a whole.

3.2 Folk Classification of the Landscape

This chapter explains the diversity of the ecological conditions and the landscape of the area under study. Previous forest ecological and floristic surveys (Whitmore 1966) and plant resource surveys (Henderson and Hancock 1988) provide information that allows researchers to understand the variation of the landscape and the vegetation of the study area; these surveys cover the entire country known as Solomon Islands and include ethnoecological studies in the neighboring Marovo region (Hviding 1996, 2005) as well as providing brief descriptions of the Roviana landscape (Aswani and Vaccaro 2008). These studies show how the local names people use to describe different ecological conditions vary from place to place. For example, Hviding and Bayliss-Smith (2000) classified local forests into 15 forest types. Whitmore (1966) and Henderson and Hancock (1988) suggested that forests consist of mosaics of different land types that reflect the timing and impacts of various human activities. This chapter, based on vegetation surveys and interviews with local people, describes unique human–forest interrelationships.

In this study, locally recognized land use types were identified through interviews with four local elders who had been recommended as forest experts by a committee of local leaders. The author visited various locations with these experts, who identified the local Roviana names for different land use types, including a variety of forested types.[1] The ecological characteristics of different land use types were also determined during these interviews. As a result, 14 land use types were identified of which 12 occur on the main island of New Georgia and seven on the barrier islands (e.g., Ndora Island, an extension of Olive). The author observed and participated in various subsistence activities to confirm that these land use types were widely recognized and frequently referred to in daily life.

Figure 3.1 shows the vegetation and land use types as recognized by the local people (also see Fig. 2.2 for the current geographical setting). In the Roviana language, forest is generally called *hinqohinqo* (or *soloso* for deep inner forest), but 12 names were used to describe various forest vegetation types. They were primary

[1] In our protocol, any disagreements about the names of specific land use types were to be resolved by discussion among the experts, authors, and other villagers, although such disagreements rarely occurred. Although only four experts were consulted, the classifications reflected widespread recognition of land types by the villagers.

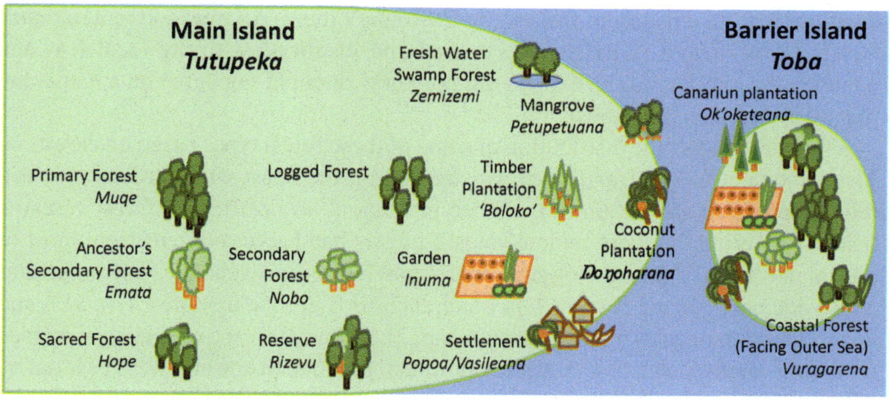

Fig. 3.1 Forest and land use classifications in the local Roviana language

forest (*muqe*), ancestor's secondary forest (*emata*), sacred forest (*hope*), secondary forest (*nobo*), reserve (*rizevu*), logged forest (no Roviana term), timber plantation (*boloko*), fresh water swamp forest (*zemizemi*), mangrove (*petupetuana*), coastal forest (facing outer sea) (*vuragarena*), coconut plantation (*ŋoŋoharana*), and canarium plantation (*ok'oketeana*). Additionally, two less forested land use types included settlement (*popoa* or *vasileana*) and garden (*inuma*). The local people call the geographic setting of New Georgia Island *tutupeka* and that of the barrier islands *toba* and understand that the soil and vegetation differ on the barrier islands when compared with those of the main island.

To understand the similarity and dissimilarity of various forest types, and the related forest biodiversity, quadrat vegetation surveys were conducted in seven forest types: primary forest (*muqe*) on the (1) main island and (2) barrier island, secondary forest on the main island ((3) *nobo*, 15-year-old fallow land) and on the barrier island ((4) 15-year-old fallow land), (5) reserve forest (*rizevu*) on the main island, (6) selectively and commercially logged forest (originally primary forest; 8–9 years after operation) on the main island, and (7) mangrove forest (*petupetuana*). The ancestor's secondary (*emata*) and sacred (*hope*) forests existed as patches in primary or secondary forests. These were excluded from the survey because local experts found these forest types difficult to distinguish independently and partly because of their small size. Four 25 × 25-m quadrats were established in each forest type, i.e., 0.25 ha in total per forest type.[2] Because the local people near the rural village recognized two separate reserve forest areas, four quadrats (two in each of these two forests) were established: i.e., 0.5 ha in total for this forest type only. For the statistical analysis, the total number of trees counted in the reserve forests was halved so that all forest types were comparable. The vegetation surveys were

[2] Small plots can been problematic in vegetation surveys; however, quadrats of 0.25 ha or less have been successfully used when locally defined forest types were not extensive enough to establish larger plots.

conducted in the territory of the Saikile Customary Land. All trees > 10 cm in diameter at breast height (DBH) were counted and identified with the local Roviana name by the experts.[3] If the experts did not know or could not agree on a name, the plant was listed as "unknown."

Figure 3.2 shows the distribution of some of these forest types based on elevation. The settlement (*popoa/vasileana*) was located on the coast with gardens (*inuma*), fallow secondary forest (*nobo*), timber plantations (*boloko*), and forest reserves (*rizevu*) covering the more remote lands. Loggers had harvested trees from most of the inland areas, although logging operations had been excluded from *tabu ples* which included sacred forest (*hope*) and ancestor's secondary forest (*emata*), and these areas remained as mature forest (Chap. 2). Primary forest (*muqe*) only occurred far inland. Barrier islands served as a traditional production site and were covered by gardens (*inuma*), fallow secondary forest (*nobo*), primary forest (*muqe*), and *Canarium* nut plantations (*ok'oketeana*). The *Canarium* nut plantations, which were recognized as property of the ancestral chiefs, were not frequently observed and were mostly regarded as untouchable forest on the main island. Many coastal zones were covered by coconut plantations (*ŋoŋoharana*). Figure 3.2 does not show swamp forest (*zemizemi*) and mangrove (*petupetuana*) located in coastal or lowland areas.

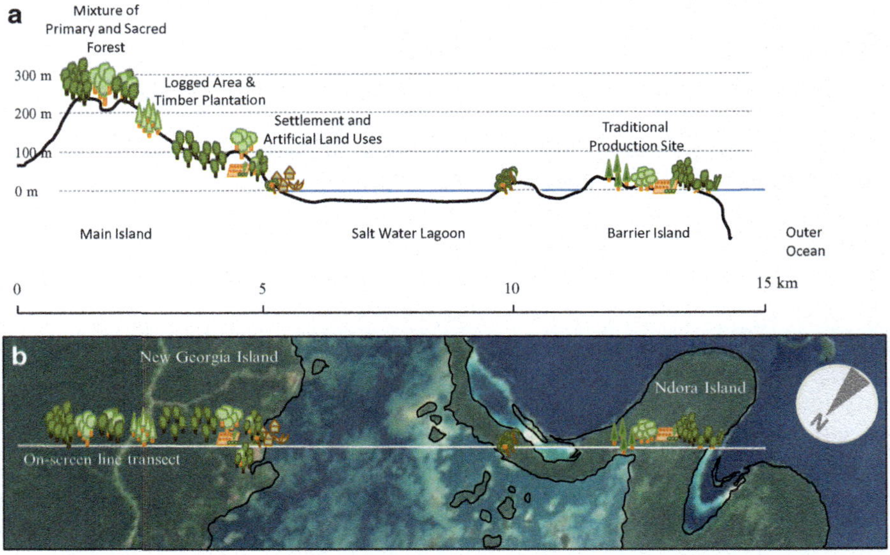

Fig. 3.2 *Vertical* (**a**) and *horizontal* (**b**) distribution of different land use types. A sectional view (Fig. 3.2a) was drawn based on a digital elevation model (USGS 2004)

[3] Botanical specimens were collected, stored in liquid alcohol in the field, and later dried at Munda Forestry Station, Ministry of Forestry of Solomon Islands Government, so that the author took these specimens to be identified. Specimens were housed at BSIP.

3.3 Primary and Secondary Forest

This section explains the detailed ecological and social roles of the land use types shown in Figs. 3.1 and 3.2. First, primary forests are called *muqe* and are those forests recognized as being unmodified by humans. In reality, people collect non-timber forest products or hunt wild pigs in the *muqe*; these forests are not truly pristine but rather experience human–forest interactions. Additionally, Bayliss-Smith and others (2003) reported that "primary" forests on New Georgia Island may actually include sites of former settlements, forest clearings, and agricultural fields (irrigated taro terraces called *ruta*).

Survey reports from previous geological and vegetation studies characterized the land types present in New Georgia by the dominant canopy species present. For example, lowland forests are typed using *Calophyllum vitiense*, *C. kajewskii*, *Dillenia* spp., and *Campnosperma brevipetiolata*. Furthermore, *Terminalia brassii* occurs in swamps, while *Pometia pinnata* and *Vitex cofassus* are dominant on the barrier islands, based on my field observations and a geological report (Wall and Hansell 1975). Table 3.1 lists the trees found in this study in quadrats for primary forest (*muqe*) in the main island (*tutupeka*). Trees are listed in sequence by the number of trees present in the quadrats, regardless of the canopy size or tree height. Therefore, smaller undergrowth trees apparently accounted for a large portion of the vegetation; however, *Calophyllum* spp. (*bunibuni* and *horehore*) formed the main canopy as reported previously. Table 3.2 showed the trees found in the primary forest of the barrier island; this type of forest was characterized by the presence of large trees such as *P. pinnata* (*qema*) and *V. cofassus* (*vasara*), while no *Calophyllum* spp. occurred there.

One type of old secondary forest is called *emata*. Local experts describe this forest type as an ancestor's secondary forest because previous generations of people had deforested and abandoned the land before the current people were born. According to the interviewees, such forests typically have some gigantic trees with high customary value, such as ngali nut (*tovinia: Canarium salomonense*), or archaeological taro terrace sites (*ruta*) exist nearby (Fig. 3.3). The *emata* is sometimes recognized as the estate or property of a group or individuals providing evidence of their ancestor's customary ownership of that land (Aswani 2000; Hviding and Bayliss-Smith 2000). Meanwhile, forest areas previously inhabited by ancestors or used for rituals are now considered shrines (*hope*) and where entry and tree felling are prohibited (Fig. 3.4). The people believe supernatural powers (*tomate*) will punish trespassers. These sacred forests represent an aspect of the people's customary land management.

The word *nobo* identifies fallow or secondary forests that had been abandoned after shifting cultivation. According to the interviewees, the residents gather medicinal plants, small trees, and palms in the *nobo* and use them to build houses in other locations. During our vegetation surveys, *Trichospermum incanum* (*pusipusi gurapa*) and *Macaranga* species (*tobo* and *tukituki*) dominated secondary forest (Table 3.3). Additionally, the forests on the main island characteristically had

Table 3.1 Trees (>10 cm diameter at breast height found in 1/4 ha quadrats analyzed in primary forest (*muqe*) on the main island (*tutupeka*)

Scientific name	Roviana name	No. in 1/4 ha quadrats
Belliolum burttianum A. C. Sm.	Unknown	37
Aglaia brassii Merr. and L. M. Perry	*Kokeqolo*	19
Calophyllum paludosum C. T. White	*Bunibuni*	10
Garcinia celebica L., *G. solomonensis* A. C. Sm.	*Gaekubo*	10
Haplolobus floribundus (K. Schum.) H. J. Lam	*Tila*	10
Syzygium onesimum Merr. and L. M. Perry	*Pidiki*	9
Melastoma sp.	*Romo*	8
Syzygium sp.	*Hioko soloso*	7
Gonystylus macrocarpus C. T. White, *G. macrophyllus* (Miq.) Airy Shaw	*Valo*	7
Neoscortechinia forbesii (Hook. f.) C. T. White	*Bolava*	6
Syzygium sp.	*Hioko*	6
Weinmannia blumei Planch.	*One*	4
Unidentified	*Bebeu*	3
Diospyros ferrea (Willd.) Bakh.	*Kureu*	3
Timonius uniflorus (Banks ex c. F. Gaertn.) Govaerts	*Suri*	3
Unidentified	Unknown	3
Codiaeum variegatum (L.) Rumph. ex A. Juss., *Horsfieldia solomonensis* A. C. Sm	*Zazalagaba*	3
Smilax zeylanica L.	*Arosopatu*	2
Gomphandra montana (G. Schellenb.) Sleumer	*Hapehape baruku*	2
Hydriastele macrospadix (Burret) W.J. Baker and Loo	*Kaŋana*	2
Micromelum pubescens Blume	*Lomalomata*	2
Unidentified	Unknown	2
Unidentified	*Vao*	2
Horsfieldia spicata (Roxb.) J. Sinclair	*E'ehara*	1
Calophyllum kajewskii A. C. Sm.	*Horehore*	1
Amoora cucullata Roxb.	*Lulua*	1
Parinari salomonensis C. T. White	*Malutita*	1
Podocarpus neriifolius D. Don	*Mou*	1
Parinari glaberrima Hassk.	*Tita*	1
Syzygium sp.	*Tivativa*	1
Schefflera stahliana (Warb.) Frodin	*Vokuvoku*	1

several *Commersonia bartramia* (*zamara*) plants and a pioneer species, white beech (*qoliti*: *Gmelina moluccana*). Currently, white beech is the only tree used for canoe building, making it a key species in this book and for the local people. *Syzygium* spp. (*hioko*) and *Flueggea flexuosa* (*mavuana*) were common on the barrier islands

Table 3.2 Trees (>10 cm diameter at breast height found in 1/4 ha quadrats analyzed in primary forest (*muqe*) on the barrier island (*toba*)

Scientific name	Roviana name	No. in 1/4 ha quadrat
Aglaia brassii Merr. and L. M. Perry	*Kokeqolo*	38
Glochidion sp.	*Puza*	16
Oxymitra macrantha Hemsl.	*Vola*	13
Syzygium onesimum Merr. and L. M. Perry	*Pidiki*	12
Horsfieldia spicata (Roxb.) J. Sinclair	*E'ehara*	6
Vitex cofassus Reinw. ex Blume	*Vasara*	6
Pometia pinnata J. R. Forst. and G. Forst.	*Qema*	5
Neonauclea spp.	*Vosevose*	4
Diospyros ellipticifolia Bakh., *D. hebecarpa* A. Cunn. ex Benth.	*Gaemuho*	3
Syzygium sp.	*Hiokomasa*	3
Medusanthera papuana (Becc.) R. A. Howard	*Garugarumu*	2
Geniostoma rupestre J. R. Forst. and G. Forst.	*Vasavasara*	2
Syzygium sp.	*Haila*	1
Ficus spp.	*Kalala*	1
Unidentified	*Kosokosohabili*	1
Gnetum latifolium Blume	*Liqeliqe*	1
Amoora cucullata Roxb.	*Lulua*	1
Horsfieldia polyantha Warb.	*Pike*	1
Alstonia spectabilis R. Br.	*Totuana*	1

(Table 3.4). The villagers sometimes specifically used the name *Hiq'hiqo*, a general term for forest, to mean the old secondary forest. Villagers sometimes hunt pigs in primary and secondary forests.

3.4 Logged Forest and Timber Plantations

Logged forest is now a prevalent forest type on the island of New Georgia, although it has no Roviana name. The major timber targets are upper canopy species with commercial value, such as *Calophyllum* spp. (*buni* or *bunibuni*), *P. pinnata* (*qema*), *Dillenia salomonensis* (*kapuhu*), *T. brassii* (*pepeo*), and *V. cofassus* (*vasara*). Some species, such as white beech, are protected by agreements between the local people and logging companies. Additionally, forestry policies and laws prohibit logging near rivers, flowing streams, and on steep hills as well as the felling of trees of<60 cm DBH. Table 3.5 shows the different trees found in the logged forest on the main island. Typically, pioneer species such as *Macaranga* spp. (*tobo* or *tuki-tuki*) grew with residual primary forest species (e.g., *Calophyllum* spp.) that remained after tree harvesting.

Fig. 3.3 Wild *Alocasia macrorrhiza* grown in an archaeological taro terrace site (*ruta*) near Kozo in 2002

Fig. 3.4 An elder collected and carried rattan near sacred forest (*hope*). Villagers were allowed to collect for subsistence even near the sacred forest (Photo taken at Olive in 2009)

Table 3.3 Tree species (>10 cm diameter at breast height) found in 1/4 ha quadrats analyzed in secondary forest (*nobo*) on the main island (*tutupeka*)

Scientific name	Roviana name	No. in 1/4 ha quadrat
Trichospermum incanum Merr. and L. M. Perry	*Pusipusi gurapa*	94
Commersonia bartramia (L.) Merr.	*Zamara*	17
Terminalia calamansanay Rolfe	*Nabinabisi*	7
Macaranga lanceolata Pax and K. Hoffm.	*Tobo*	7
Euodia elleryana F. Muell.	*Bosi suka*	6
Macaranga spp.	*Tukituki*	6
Pometia pinnata J. R. Forst. and G. Forst.	*Qema*	4
Euodia solomonensis Merr. and L. M. Perry	*Bosi*	3
Semecarpus forstenii Blume	*Sakita*	3
Garcinia celebica L., *G. solomonensis* A. C. Sm.	*Gaekubo*	2
Dillenia ingens B. L. Burtt	*Hebere*	2
Unidentified	*Kosokosohabili*	2
Unidentified	*Pusipusi kaqele*	2
Macaranga gigantea Müll. Arg.	*Rarabuta*	2
Alstonia spectabilis R. Br.	*Totuana*	2
Gonystylus macrocarpus C.T. White, *G. macrophyllus* (Miq.) Airy Shaw	*Valo*	2
Albizia falcataria (L.) Fosberg	*Bue*	1
Trema orientalis (L.) Blume	*Dodoru*	1
Ficus imbricata Corner	*Duvi*	1
Syzygium sp.	*Hioko*	1
Calophyllum kajewskii A. C. Sm.	*Horehore*	1
Gulubia macrospadix (Burret) H. E. Moore	*Kaŋana*	1
Elaeocarpus floribundus Blume	*Konu*	1
Micromelum pubescens Blume	*Lomalomata*	1
Parinari salomonensis C. T. White	*Malutita*	1
Rhus taitensis Guill.	*Natoŋo*	1
Canarium vitiense A. Gray	*Okokete*	1
Phyllanthus ciccoides Mull. Arg.	*Pakopako*	1
Litsea sp.	*Petepete kaqele*	1
Glochidion sp.	*Puza*	1
Gmelina moluccana Backer ex K. Heyne	*Qoliti*	1
Unidentified	Unknown	1
Syzygium cinctum Merr. and L. M. Perry, *S. aqueum* (Burm. f.) Alston	*Uvolo*	1
Unidentified	*Vaosoloso*	1
Geniostoma rupestre J. R. Forst. and G. Forst.	*Vasavasara*	1
Fagraea racemosa Jack	*Zelemuru*	1

Table 3.4 Trees (>10 cm diameter at breast height) found in 1/4 ha quadrats analyzed in secondary forest (*nobo*) on the barrier island (*toba*)

Scientific name	Roviana name	No. in 1/4 ha quadrat
Macaranga spp.	*Tukituki*	26
Syzygium sp.	*Hioko*	10
Oxymitra macrantha Hemsl.	*Vola*	10
Horsfieldia spicata (Roxb.) J. Sinclair	*E'ehara*	7
Amoora cucullata Roxb.	*Lulua*	7
Horsfieldia polyantha Warb.	*Pike*	6
Flueggea flexuosa Müll. Arg.	*Mavuana*	5
Vitex cofassus Reinw. ex Blume	*Vasara*	5
Medusanthera papuana (Becc.) R. A. Howard	*Garugarumu*	4
Gnetum gnemon L.	*Leqe*	4
Glochidion sp.	*Puza*	4
Semecarpus forstenii Blume	*Sakita*	4
Euodia solomonensis Merr. and L. M. Perry	*Bosi*	3
Ficus septica Burm. f.	*Huhu*	3
Diospyros ellipticifolia Bakh., *D. hebecarpa* A. Cunn. ex Benth.	*Gaemuho*	2
Unidentified	*Halagire*	2
Unidentifieid	*Kosokosohabili*	2
Macaranga lanceolata Pax and K. Hoffm.	*Tobo*	2
Dysoxylum excelsum Blume	*Vuagore*	2
Ficus imbricata Corner	*Duvi*	1
Unidentified	*Hirihiri*	1
Aglaia brassii Merr. and L. M. Perry	*Kokeqolo*	1
Sterculia shillinglawii F. Muell.	*Kokolomana*	1
Caryota rumphiana Mart.	*Piku*	1
Ficus lancibracteata Corner	*Rapa*	1
Timonius forsteri DC.	*Suri*	1
Alstonia scholaris (L.) R. Br.	*Taŋovo*	1
Haplolobus sp.	*Turutono*	1
Phyllanthus sp.	Unknown	1
Elaeocarpus miegei Weibel	Unknown	1
Unidentified	Unknown	1
Syzygium cinctum Merr. and L. M. Perry, *S. aqueum* Alston	*Uvolo*	1
Hibiscus tiliaceus L.	*Varu*	1
Neonauclea spp.	*Vosevose*	1
Fagraea racemosa Jack	*Zelemuru*	1

Table 3.5 Trees (>10 cm diameter at breast height) found in 1/4 ha quadrats analyzed in logged forest

Scientific name	Roviana name	No. in 1/4 ha quadrat
Haplolobus floribundus (K. Schum.) H. J. Lam	*Tila*	22
Campnosperma brevipetiolata Volkens	*Pokopoko*	21
Palaquium erythrospermum H. J. Lam	*Paloto*	10
Gonystylus macrocarpus C. T. White, *G. macrophyllus* (Miq.) Airy Shaw	*Valo*	10
Neoscortechinia forbesii (Hook. f.) C. T. White	*Bolava*	9
Garcinia celebica L., *G. solomonensis* A. C. Sm.	*Gaekubo*	8
Codiaeum variegatum Blume, *Horsfieldia solomonensis* A. C. Sm	*Zazalagaba*	8
Macaranga lanceolata Pax and K. Hoffm.	*Tobo*	6
Syzygium sp.	*Zuapa*	6
Elaeocarpus floribundus Blume	*Konu*	5
Syzygium onesimum Merr. and L. M. Perry	*Pidiki*	5
Medinilla mortonii Hemsl.	Unknown	5
Unidentified	*Hirihiri*	4
Aglaia brassii Merr. and L. M. Perry	*Kokeqolo*	4
Glochidion sp.	*Puza*	4
Timonius uniflorus (Banks ex C. F. Gaertn.) Govaerts	*Suri*	4
Calophyllum paludosum C. T. White	*Bunibuni*	3
Trichospermum incanum Merr. and L. M. Perry	*Pusipusi gurapa*	3
Parinari glaberrima Hassk.	*Tita*	3
Haplolobus sp.	*Turutoŋoro*	3
Melastoma polyanthum Blume	*Vovou*	3
Horsfieldia spicata (Roxb.) J. Sinclair	*E'ehara*	2
Diospyros ellipticifolia Bakh., *D. hebecarpa* A. Cunn. ex Benth	*Gaemuho*	2
Unidentified	*Kosokosohabili*	2
Weinmannia blumei Planch.	*One*	2
Litsea sp.	*Petepetekaqele*	2
Macaranga spp.	*Tukituki*	2
Syzygium cinctum Merr. and L. M. Perry, *S. aqueum* Alston	*Uvolo*	2
Horsfieldia irya (Gaertn.) Warb.	*Babageva soloso*	1
Ficus imbricata Corner	*Duvi*	1
Maranthes corymbosa Blume	*Giza*	1
Dillenia ingens B. L. Burtt	*Hebere*	1
Unidentified	*Hirihir soloso*	1
Calophyllum kajewskii A. C. Sm.	*Horehore*	1
Ficus septica Burm. f.	*Huhu*	1
Endospermum formicarum Becc.	*Kakadikana*	1

(continued)

Table 3.5 (continued)

Scientific name	Roviana name	No. in 1/4 ha quadrat
Gulubia macrospadix (Burret) H. E. Moore	*Kaŋana*	1
Dillenia salomonensis (C. T. White) Hoogland	*Kapuhu*	1
Cyathea spp.	*Kuqui*	1
Unidentified	*Levaleva*	1
Unidentified	*Likelike*	1
Gnetum latifolium Blume	*Liqeliqe*	1
Micromelum pubescens Blume	*Lomalomata*	1
Terminalia calamansanay Rolfe	*Nabinabisi*	1
Horsfieldia polyantha Warb.	*Pike*	1
Macaranga gigantea Müll. Arg.	*Rarabuta*	1
Macaranga quadriglandulosa Warb.	*Rarabuta soloso*	1
Unidentified	Unknown	1

Although the reforestation of timber plantations commenced in the late 1990s, this activity intensified in the 2000s (Furusawa et al. 2004; Racelis and Aswani 2011). Many logged forests were clear-cut to create lands for timber plantations. Such community-level reforestation has accelerated under the leadership of the large church group, the Christian Fellowship Church, on New Georgia Island. Teak (*Tectona grandis*) and rainbow eucalyptus (*Eucalyptus deglupta*) are common plantation species. Additionally, oil palm (*Elaeis guineensis*) plantations have been expanded in the northern part of New Georgia; however, this has occurred only to a limited extent in Roviana during the study period. Although large-scale logging operations are not conducted in the urban territory, expectations of future cash income have driven the people to clear-cut the secondary forest near urban areas to convert them to timber plantations.

3.5 Forest Reserve

Reserve forests, or *rizevu* (borrowed from the English word "reserve"), grow near the Olive village settlement. These nature reserves were established in accordance with the recommendations of local chiefs and Christian Fellowship Church leaders, and these leaders prohibit logging by multinational companies and agricultural cultivation in these areas. However, the local people are allowed to harvest trees for personal use and to remove several non-timber forest products (e.g., *Calophyllum* spp. and *Dillenia* spp. as materials for tools, buildings, medicine, and fuel). Therefore, pioneer species grew in the gaps created by human activities. The vegetation in these reserve forests differs from that in both primary and secondary forests because of these continuous anthropogenic impacts (Table 3.6; see detailed analyses

Table 3.6 Trees (>10 cm diameter at breast height) found in 1/2 ha quadrats (sum of two 1/4 quadrats) made in the reserve (*rizevu*)

Scientific name	Roviana name	No. in 1/2 ha quadrat
Garcinia celebica L., *G. solomonensis* A. C. Sm.	*Gaekubo*	38
Parinari glaberrima Hassk.	*Tita*	18
Haplolobus floribundus (K. Schum.) H. J. Lam	*Tila*	17
Timonius forsteri DC.	*Suri*	16
Neoscortechinia forbesii (Hook. f.) C. T. White	*Bolava*	12
Glochidion sp.	*Puza*	12
Calophyllum paludosum C. T. White	*Bunibuni*	10
Calophyllum kajewskii A. C. Sm.	*Horehore*	9
Aglaia brassii Merr. and L. M. Perry	*Kokeqolo*	8
Gonystylus megacarpus C. T. White, *G. macrophyllus* (Miq.) Airy Shaw	*Valo*	8
Elaeocarpus floribundus Blume	*Konu*	7
Horsfieldia spicata (Roxb.) J. Sinclair	*E'ehara*	6
Dillenia salomonensis (C. T. White) Hoogland	*Kapuhu*	6
Amoora cucullata Roxb.	*Lulua*	6
Casearia grewiifolia Vent.	Unknown	6
Maranthes corymbosa Blume	*Giza*	5
Campnosperma brevipetiolata Volkens	*Pokopoko*	5
Unidentified	*Bebeu*	3
Diospyros ferrea (Willd.) Bakh.	*Kureu*	3
Terminalia calamansanay Rolfe	*Nabinabisi*	3
Canarium vitiense A. Gray	*Okokete*	3
Triumfetta procumbens G. Forst.	Unknown	3
Euodia elleryana F. Muell.	*Bosi suka*	2
Ficus imbricata Corner	*Duvi*	2
Syzygium sp.	*Hiokosoloso*	2
Unidentified	*Hirihiri*	2
Weinmannia blumei Planch.	*One*	2
Palaquium erythrospermum H. J. Lam	*Paloto*	2
Macaranga lanceolata Pax & K. Hoffm.	*Tobo*	2
Litsea sp.	Unknown	2
Horsfieldia irya (Gaertn.) Warb.	*Babageva soloso*	1
Euodia solomonensis Merr. and L. M. Perry	*Bosi*	1
Diospyros ellipticifolia Bakh., *D. hebecarpa* A. Cunn. ex Benth	*Gaemuho*	1
Calophyllum sp.	*Hame*	1
Dillenia ingens B. L. Burtt	*Hebere*	1
Syzygium sp.	*Hioko*	1
Syzygium sp.	*Hioko masa*	1

(continued)

Table 3.6 (continued)

Scientific name	Roviana name	No. in 1/2 ha quadrat
Syzygium sp.	*Hioko muqe*	1
Gulubia macrospadix (Burret) H. E. Moore	*Kaŋana*	1
Syzygium onesimum Merr. and L. M. Perry	*Pidiki*	1
Pometia pinnata J. R. Forst. and G. Forst.	*Qema*	1
Ficus lancibracteata Corner	*Rapa*	1
Semecarpus forstenii Blume	*Sakita*	1
Haplolobus sp.	*Turutonoro*	1
Stemonurus celebicus Valeton ex Koord.	Unknown	1
Schizomeria brassii Mattf.	Unknown	1
Unidentified	*Vao*	1
Unidentified	*Vao soloso*	1
Fagraea racemosa Jack	*Zelemuru*	1

in the next section). A mixture of climax and pioneer species characterize these forests. For example, people frequently collect the fallen fruit of *Parinari glaberrima* (*tita*), one such pioneer species (Fig. 3.5). Its fruit is shredded or pounded to make putty for repairing holes and cracks in canoes. Urban areas do not have reserve forests.

3.6 Mangrove

Mangroves (*petupetuana*) play an important role for the rural people by providing not only plant resources but also hunting grounds for shellfish and crabs. For example, the interviewees stated that on one occasion, a group of people protested when an inflow of red soil decreased the populations of shellfish and crabs in local mangroves, and their actions forced a logging company to halt operations. Table 3.7 shows that *Bruguiera gymnorrhiza* (*petu*) dominated the mangroves, followed by *Rhizophora apiculata* (*petu roga*). In contrast, mangroves are quite rare in urban villages because the coastal areas have been converted to infrastructure and settlements.

3.7 Horticultural Gardens and Settlements

Sweet potatoes (*luzu vaka*: *Ipomoea batatas*) and cassava (*ore marihi*: *Manihot esculenta*) as well as traditional root crops, such as taro (*talo*: *Colocasia esculenta*) and a variety of yams (*marihi*, *pana*, etc.: *Dioscorea* spp.) serve as the primary

Fig. 3.5 *Parinari glaberrima* (tita tree) collected in the forest reserve; villagers use the sticky, oily fruit as putty for sealing cracks and holes in dugout canoes (Photo taken at Olive in 2001)

Table 3.7 Trees (>10 cm diameter at breast height) found in 1/2 ha quadrats analyzed in mangroves (*petupetuana*)

Scientific name	Roviana name	No. in 1/4 ha quadrat
Bruguiera gymnorrhiza (L.) Lam.	*Petu*	75
Rhizophora apiculata Blume	*Petu roga*	30
Ceriops tagal (Perr.) C. B. Rob.	*Vorusu*	13
Xylocarpus granatum J. Koenig	*Koe*	5
Sonneratia caseolaris (L.) Engl.	*Tototu*	4
Heritiera littoralis Aiton	*Babageva*	3
Cordia subcordata Lam.	*Naqinaqi*	3
Pandanus sp.	*Agana*	2
Cocos nucifera L.	*Ɖohara*	1
Dolichandrone spathacea K. Schum.	*Tui*	1

Fig. 3.6 Various useful plants growing in a settlement, Olive, in 2003

cultivated crops; villagers also plant these in roughly half of the gardens. Nonedible plants, such as *Coleus* spp., are also planted as ornamentals or for magic (e.g., protecting crops from pest animals) in gardens. Settlements are called *popoa* or *vasileana*. Almost all trees and palms growing in the settlements were either domesticated or semidomesticated and were used for various purposes. For example, all trees of > 10 cm DBH found along a road in Olive village were either coconut (*ŋohara*: *Cocos nucifera*), betel nut (*heta*: *Areca catechu*), cut nut (*kinu*: *Barringtonia procera*), tropical almond (*tatalise*: *Terminalia catappa*), or kapok (*lozi*: *Ceiba pentandra* and *Bombax malabaricum*) (Fig. 3.6). Villagers planted some introduced flowering plants, such as *Catharanthus roseus*, in some houses for ornamental purposes.

3.8 Similarity and Dissimilarity Between Forests

The discussion above describes and explains different land uses. Field surveys recorded 168 (31 species), 120 (49), 181 (48), and 180 (49) individual trees (>10 cm DBH) in the four quadrats in the primary, reserve, secondary, and logged forests, respectively, on the main island (Tables 3.1, 3.3, 3.5, and 3.6). Additionally, 137 trees (10 species) were observed in the mangrove habitat (Table 3.7). On the barrier

Table 3.8 Sørensen–Dice similarity index (%) used to compare the similarity of different forest types

	Main island				Barrier island	
	Reserve	Secondary	Mangrove	Logged	Primary	Secondary
Main						
Primary	49.5	27.8	0	40.0	16.0	14.9
Reserve		45.9	0	52.6	17.4	33.8
Secondary			0	49.0	20.9	33.3
Mangrove				0	0	0
Logged (selectively)					21.9	31.1
Barrier						
Primary						40.0

island, 117 (19 species) and 124 (36) trees were found in the primary and secondary forests, respectively (Tables 3.2 and 3.4). Table 3.8 shows the similarities between the different types of vegetation.[4] The primary forests on the main and barrier islands were only half as similar (16.0 %) as the respective secondary forests (33.3 %). Vegetation in the reserve forests was 45.9–52.6 % similar to that found in the primary, secondary, and logged forests on the main island. Species found in the mangrove forest were not observed in other forest types.

In summary, these findings suggest that each forest type represented a different vegetative community, with relatively low rates of similarity between communities. While primary forest represented areas with few human impacts, several unique species were also found in human-modified forests (i.e., reserve forests and secondary forests). The vegetation varied markedly between the main and barrier islands. Thus, each forest type contained a different type of biodiversity and this level of diversity is related to human activities. Historically, human populations have also affected "primary forest." People have clearly created the existing secondary forest and reserve forests. Nevertheless, these forests have a unique level of diversity not found in gardens and settlements. Additionally, these various forest types contribute to the overall landscape diversity of the study area.

[4] The similarity of species composition between each pair of forest types was calculated using the Sørensen–Dice similarity index:

$$\text{Sørensen–Dice similarity index (\%)} = 2c/(a+b) \times 100\ \%$$

where c is the number of species observed in both forest types and a and b are the number of species in forest types A and B, respectively.

Chapter 4
Subsistence on the Main Island, Barrier Islands, and at Sea

Abstract This chapter shows how people use different ecosystems, specifically the main island and barrier island ecosystems of the study area with their independent geological histories, that are separated by a coral lagoon sea. This chapter addresses a variety of disciplines such as land use, shifting cycles, land tenure, soil nutrients, subsistence production, vegetation, land cover change, and carrying capacity. As the results, the productivity per shifting cycle of the barrier island was the highest among the different geographical locations analyzed here. This cycle was supported by natural characteristics and social system. The main island was also useful for the villagers for conducting new economic activities such as planting of perennial cash crops rather than for subsistence production. The different uses of the barrier and main islands allowed the villagers to integrate new cash crops without the concerns of food security while receiving the benefits of sustainable production from the barrier island. The GPS tracking of villagers showed they also used a wide area of sea and a variety of marine ecosystems by traveling more than 30 km before returning home. The villagers' subsistence lifestyle depended on the diversity of the landscape/seascape and the availability of various species.

Keywords Main island • Barrier island • Carrying capacity • Sustainable production • Horticultural production • Spatiotemporal change • Marine resources • GPS tracking

4.1 Human Ecological Approach

Human ecology mainly studies and analyzes subsistence lifestyles. A number of early human ecological studies were primarily based on ecosystem approaches, where human societies were analyzed as populations within ecosystems so as to analyze the subsistence lifestyle of humans within the ecosystem. The survival strategies used for production and consumption in given ecosystems have been studied, such as the use of space and time in a subsistence society, the carrying capacity of an ecosystem in relation to domesticated animals (e.g., pig in Papua New Guinea), and food webs and energy flows in human-dominated ecosystems (Dwyer 1990; Ellen 1982; Morren 1986; Ohtsuka 1983; Sillitoe 1996). Additionally, the carrying

T. Furusawa, *Living with Biodiversity in an Island Ecosystem*, Ethnobiology,
DOI 10.1007/978-981-287-904-2_4

capacity of the limited land and space available on small islands had also been a central concern in human ecology (Bayliss-Smith 1974; Carroll 1975; Kirch and Rallu 2007). This approach is conceptually related to the background of the Darwinian theory of the evolution of species.

4.2 Subsistence Pattern

As described above, the main mode of subsistence in Roviana involved horticulture of root crops and fishing. Marine resources provided the main source of cash income. In the village of Olive, the villagers generally tended to engage in gardening on Mondays and Fridays and in fishing on Saturdays; only a few villagers engaged in hunting wild pigs on Saturdays. Tuesdays through Thursdays were spent doing community work that was designed to provide communal benefits. Village leaders chose specific daily activities for community work, e.g., repairing the primary school, fund raising for the community, etc. The villagers usually woke up around sunrise and made a fire in the kitchen area. They would go to their gardens after eating breakfast and return home in the evening. However, they sometimes do gardening early in the morning to avoid the strong sunlight at midday. Villagers also went fishing during the day, although some villagers went fishing at night or early in the morning. Activities that created cash income such as the collection of *Nassarius* shells were conducted using time before or after community work, horticulture, or fishing. On Sundays, all villagers attend church services and were prohibited from engaging in any gardening or labor; they would sometimes engage in economic activities after the Sunday evening church services. The day for fishing (or infrequently, for hunting) was understood to be a day for preparing a feast after the Sunday service.

The concepts and methods of human ecology require an understanding of how a society uses space and time for foraging activities, and this has been central to debates related to human behavioral ecology (Aswani 1997; Ellen 1982; Suda 1994). A time allocation study was conducted to track the activities of all adult members aged 18 years or older in the 15 households participating in this research study (26 males and 25 females); the households were visited every 80 min from 07:00 to 20:20 every day for seven consecutive days.[1] The participant's activities were observed and recorded during every visit (see Moji and Koyama 1985; Furusawa and Ohtsuka 2009). If the participant was not at home, the members of his/her or neighboring households were asked about the participant's activities. The survey provided 3570 spot-check observations. The activities of the participants were recorded in detail during the analysis and categorized into eight classes (Table 4.1). In this table, horticulture included gardening activities, walking and canoeing from houses to gardens, and rest periods in the gardens. Males and females used an

[1] A random sample constituted 15 households in August 2003, and all the household heads and their spouses of these households agreed to participate in the study.

Table 4.1 Time spent for different activities (average number of min/day) during daytime hours (07:00–20:20)

Activity	Time consumed (min/day, during 800 min from 07:00 to 20:20)		
	Male (N=26)	Female (N=25)	Total (N=51)
Horticulture	61.0	70.4	65.6
Fishing/shell and crab collection (for self-consumption)	60.6	26.1	43.7
Cash earning activities	61.0	67.2	64.0
Firewood and forest material collection	5.2	6.4	5.8
Housekeeping activities/hygiene	107.2	238.6	171.4
Church/community activities	106.8	80.9	94.2
Visiting town/other villages	7.4	76.3	41.1
Resting/sleeping	390.6	234.1	314.2

average total of 61 min and 70.4 min, respectively, for horticulture each week. Males spent more time (60.6 min) for fishing and shell or crab collection than females (26.1 min). Both males and females spent about 1 h per day involved in activities that created cash earnings. Females spent more time for housekeeping activities than males.

4.3 A Comparison of Human Use of the Main and Barrier Islands

As explained above, people traditionally used the main island (i.e., New Georgia Island) and small remote islands differently. No studies have analyzed the roles of the two different ecosystems in food production and consumption in response to population growth and economic development. A case observed in Olive village was interesting because, although the villagers have been integrated in the market economy in terms of cash cropping, they continue to use the barrier islands for shifting cultivation in a traditional manner (Furusawa and Ohtsuka 2009). This chapter, based on a multidisciplinary approach, reports the villagers' local strategy with regard to using different geographical settings to adapt to changing sociodemographic conditions.

Figure 4.1 shows locations where trees have been felled and cultivation has begun as seen in a very-high-resolution satellite image (IKONOS); cultivated lands were observed within settlements, along rivers, on lands disturbed by logging, and in the Raqata area on Ndora Island, while the remaining lands remained largely forested.

Olive villagers, consisting of "sub-clans" that are descendants of the Saikile clan, had usufruct of lands between the Rorosi (see Rorosi cultivation site in Fig. 4.1) and Huda (Huda in the figure) rivers amounting to approximately 2000 ha in New

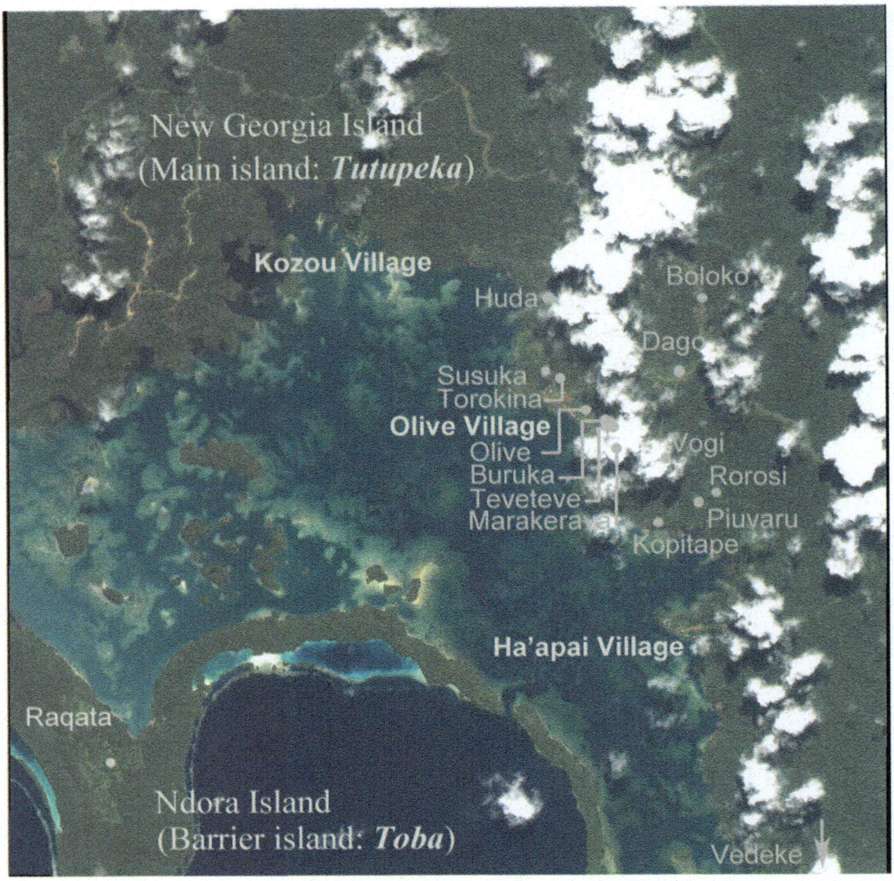

Fig. 4.1 Satellite image (IKONOS), acquired in July 2002, showing Ndora Island (the barrier island: *Toba* in the local Roviana language) and New Georgia Island (the main island: *Tutupeka*). *Gray dots* show the horticultural sites of the Olive villagers (one or more gardens at each site)

Georgia; members of a limited number of households who were descended from another sub-clan claimed ownership of areas outside these lands on the island. Figure 4.1 shows New Georgia Island where Olive villagers cultivated gardens in and around Olive settlements (local place names: Olive, Susuka, Teveteve, Torokina, Buruka, and Marakerava), the more remote lands where logging had been conducted in 1993–1994 (Dago, Boloko, and Vogi), and gardens located along rivers (Huda, Kopitape, Piuvaru, Rorosi, and Vedeke). The villagers were likely to use the logged forest because the logging road provided access to that area; therefore, this required less labor than if the villagers used forest areas that had not experienced recent disturbance. Sub-clans of Nusa Hope villagers use the Western half of Ndora Island, which covers approximately 1550 ha; the entire Saikile clan recognized approximately 760 ha of the Eastern half of the Island as communal land.

Interestingly, this part of the land—the cultivated land of the inland area called Raqata—was used communally by villagers from Olive, Ha'apai, Kozou, Nusa Hope, and several other villages.

Roviana people originally used lands to cultivate root crops. Coconut plantations in coastal areas have become a traditional form of land use for cash income since the beginning of the twentieth century. However, most of these coconut plantations have recently been less used because of the unstable market for coconuts. Similarly, cocoa was seldom sold because of the very limited market for cocoa. Reforestation of Roviana forests with imported commercial timber species such as eucalyptus and teak provides the most intense opportunities for creating cash earnings for local people since 2001. In 2001, one specific social leader was also the leader of the Christian Fellowship Church (CFC) as well as a customary chief. He encouraged reforestation in this area, although the villagers had previously mainly planted cocoa plantations; reforestation itself began at the end of the 1990s on a small scale and mainly occurred in logged areas (i.e., Dago, Boloko, and Vogi), while cocoa plantations were planted in riparian areas (e.g., Huda and Rorosi). A couple of decades will need to pass before the planted timber species can be harvested and sold.

Figure 4.2 shows a sample of typical new garden in the settlement with a variety of crops. Sweet potato serves as the main crop, while the villagers had planted cassava on the edge and taro in the middle of the gardens. As described in the Roviana language, gardening starts by clearing an undergrowth of shrubs (*pokipoki*) and by harvesting large trees (*minaho*), followed by drying and burning (*sinulu*) of the harvested shrubs and trees (*pinopa*). Then the villagers cultivate the soil and make mounds for tubers such as sweet potatoes (*piniki*) or choose appropriate locations to plant green leafy crops and other vegetables (*inililaka*), before they plant other crops (*linete*). After planting, the villagers maintain the gardens and crops (*kinopu*) by weeding until harvest (*gineligeli* for root crops and *pinakupakete* for leaves/vegetables).

The owner of the garden in Fig. 4.2 stated that he slashed and collected both trees and shrubs to burn them in a single pile; the burn sites with ashes within the garden were called *eba* (Fig. 4.3). According to him, an *eba* is not suitable for cultivating sweet potato, but cabbage and other green crops grow well there as *inililaka*. He did not slash *Canarium* nut trees that one of his parents had previously planted before that parent had abandoned the garden as fallow. Coleus (*Coleus* spp.; *boroboro* in the Roviana vernacular name) was planted to magically protect crops from pest animals.[2] Cassavas were planted at the edge of the cultivation site because this crop can grow without requiring mounds. He insisted that he would plant the timber species here after a few months; crops would be harvested here for a few more years until the planted timber species grew to shade the soil surface.

[2] *Codiaeum variegatum* (*zazalagaba*) and *Cordyline terminalis* (*zipolo*) have also been recognized as having magical uses and were planted in settlements or near gardens in Roviana and Marovo (Hviding and Bayliss-Smith 2000)

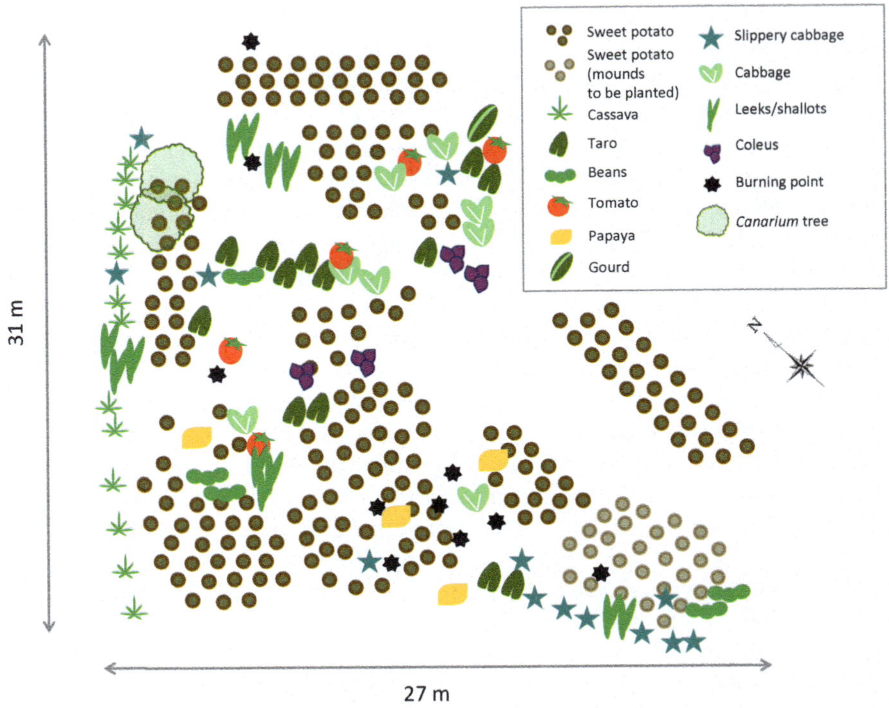

Fig. 4.2 Overview of a new settlement garden less than 1 month after clearing the forest in Torokina. Locations and sizes of crops and mounds are conceptual, not necessarily reflecting reality

Figure 4.4 shows a sample garden in the logged area. Villagers typically plant teak (*Tectona grandis*), an introduced timber species, in lines. The owner cleared the forest and then planted sweet potatoes in mounds of soil. Then he planted the above-mentioned timber species. Villagers use this common style of making a garden on the main island. Villagers might plant cassava in some gardens as the main crop along with banana and coleus for either magic or ornamental purposes.

Figure 4.5 shows a sample of a garden, the largest and one of the oldest, on Raqata, a barrier island. A variety of crops were planted using a pattern that was similar to that used for gardens on the main island. However, one important difference was that perennial cash crops (e.g., timber species) were never planted on the barrier island. In addition to main crops such as sweet potato and cassava, the villagers had planted a number of traditional staple foods such as taro, giant taro, yams, and bananas. The next sections explain why different crops were planted and different cultivation cycles were used in different areas (cultivation and fallow period).

Fig. 4.3 A villager burning slashed trees and weeds. The burning site is called an *eba* (Photo taken at Olive in 2003)

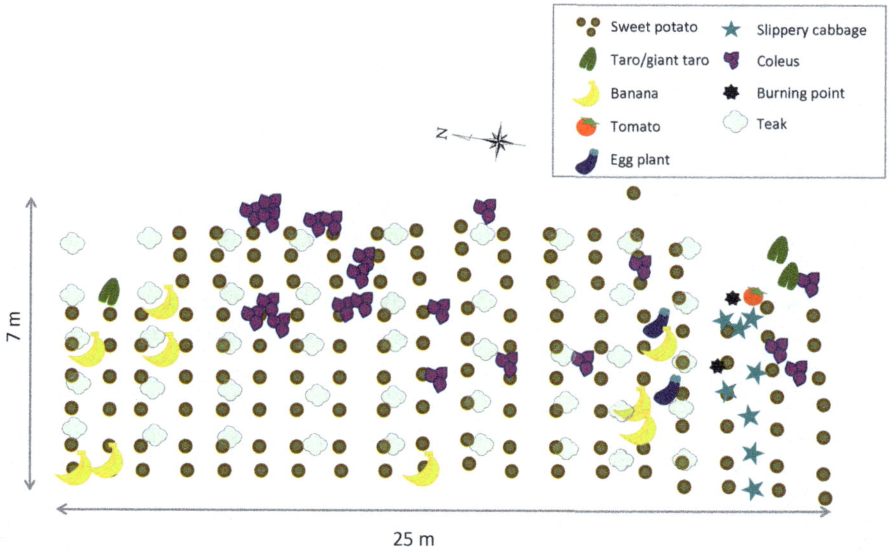

Fig. 4.4 Overview of a year-old previously logged garden site at 1.3 years after forest clearing in Vogi. Locations and sizes of crops and mounds are conceptual, not necessarily reflecting reality

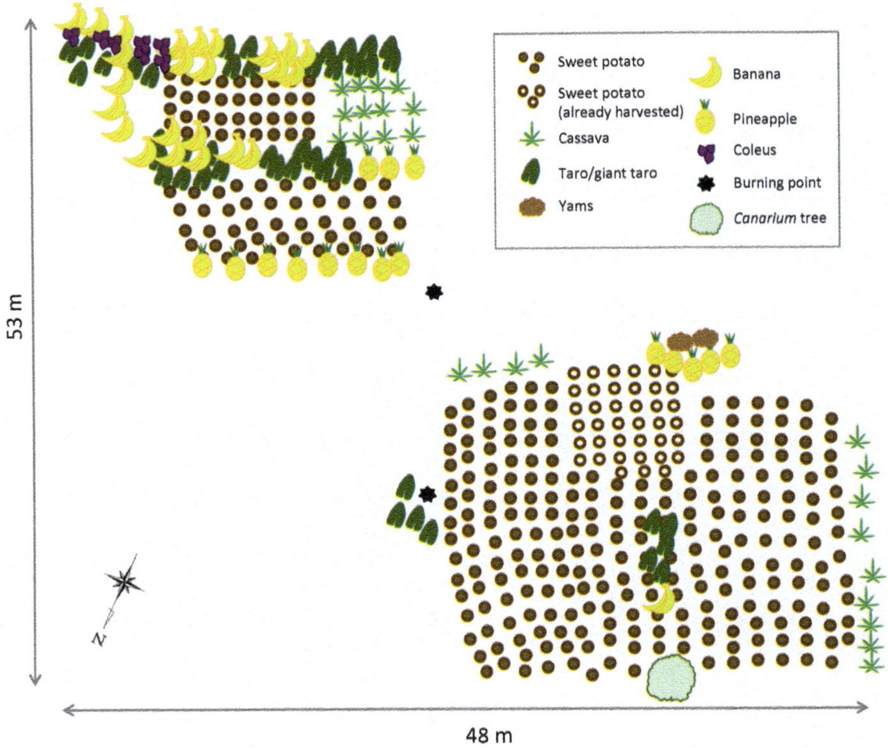

Fig. 4.5 Overview of a garden at 20 years after forest removal on Raqata, a barrier island. Locations and sizes of crops and mounds are conceptual, not necessarily reflecting reality

4.4 Spatiotemporal Distribution of Cultivation Lands

First, the timing and locations of horticultural gardens were analyzed. Table 4.2 shows the garden areas of the Olive villagers divided by locations, i.e., on the barrier island and riverside, in the settlements, and on logged lands of the main island.[3] In February 2001, 42.6 % of the horticultural land was located on the remote barrier island (Raqata area), followed by 36.5 % on riverside lands. Gardens (20.9 % of all gardens) also occupied some of the forest where selective logging had occurred in 1993–1994. No gardens were observed in the settlements. According to village

[3] A random sample constituted 14 households in February 2001 and May 2002, and 15 in August 2003 (see footnote 5), and all the household heads and their spouses of these households agreed to participate in the study. In each household, the household head and his/her spouse were asked to report the places where his/her household cultivated gardens. Subsequently, the author accompanied the participants as a part of their horticultural activities and confirmed the garden locations. The land areas of all the gardens owned by the participants were measured using a tape measure and a clinometer. Note that the villagers conducted horticultural activities throughout the year without seasonal variations, because the temperature and precipitation varied little in this region during the year.

Table 4.2 Locations and changes of garden areas of Olive villagers from 2001 to 2003

Locations	Study period (no. of participating households; total garden area)		
	2001 February (14 HHs; 2.03 ha [a])	2002 May (14 HHs; 1.95 ha)	2003 August (15 HHs; 1.38 ha)
Barrier island	42.6 %	11.9 %	34.6 %
Main island			
Riverside	36.5 %	20.3 %	15.3 %
Settlement	0 %	2.2 %	34.3 %
Logged land	20.9 %	65.6 %	15.7 %

[a]As part of the management of timber plantation species, the villagers planted seedlings of timber species in gardens; subsequently, when these trees created too much shade, the villagers moved their gardens to other locations. This caused the total garden area to be larger in 2001 or 2002 than in 2003

Fig. 4.6 A woman walking on a logging road to gardens located in previously logged areas. She holds a leaf of *Dillenia ingens* (*hebere*) as an umbrella (Photo taken at Dago in 2001)

informants, they traditionally established horticultural gardens near settlements, riversides, and barrier islands until the mid-1990s. After the logging operations, the local people extended their living area into the inner island, which had not been accessible until a road was constructed in support of the logging activities (Fig. 4.6). They also stated that they had used the areas in and around the settlements (Olive, Susuka, Marakerava, Teveteve, and Buruka) until the mid-1980s, and these places were in the fallow period, i.e., they were not in use during the study period. Cash crops such as cacao (*Theobroma cacao*) and betel nut (*Areca catechu*) were observed in riverside gardens to a small extent; these were planted in mixed gardens with the main crops that formed the villagers' diet, such as sweet potatoes and cassavas. Reforestation with commercial timber species such as eucalyptus and teak began in 1997, and continued, albeit to a limited extent, until 2001. The villagers planted seedlings of timber species in gardens. Subsequently, when the planted trees grew and began to shade the ground, the villagers would stop cultivating those sites. That is, the trees created too much shade and heavy rains would begin to cause soil erosion, even though the gardening activities were helpful for the growth of young timber/tree seedlings. In February 2001, planted timber species were observed in certain sections of the gardens in the logged land, but never in other areas.

The distribution of the cultivation area changed dramatically in 2003. In May 2002, 65.6 % of the horticultural area was located in logged lands, while the barrier island and riverside gardens contributed 11.9 % and 20.3 %, respectively, of the total cultivated land. This change was related to the recommendations of the CFC, because the CFC encouraged reforestation. All of the Olive villagers attended this church. Consequently, the village leaders decided to divide the logged forest between households in the community by allocating land to every household so they could plant their own seedlings. Almost all these lands were initially used as gardens. At this time, the villagers used the traditional garden locations less frequently and spent more time cultivating and managing crops in the logged lands than in previous periods.

In August 2003, when the planted timber tree seedlings had grown sufficiently to shade the soil surface and limit the growth of root crops, the logged forest accounted for only 15.7 % of the total cultivation area, while the cultivation area in the barrier island increased to 34.6 %. According to the villagers participating in this study, the seedlings they planted grew to cover ground surface of the logged forest area in about 2–3 years, and when that occurred, they returned to the traditional garden lands on the barrier island (Raqata) (Fig. 4.7). The extent of the cleared land along the riverside decreased, because the villagers expanded the reforestation of commercial timber species to this land. Forests in and around the settlement were largely cleared for gardens during this period because, according to the villagers, it required more than a decade of fallow for the soil fertility to recover. In summary, the barrier island was used as the main horticultural area; the cultivated land was frequently, but temporarily, disturbed by the villagers' strategies for the plantation of perennial timber species on the main island of New Georgia. The perennial crops spread from the logged lands that the villagers had not used until the 1990s to riverside lands that the villagers had traditionally used. The expansion of settlement gardens reflected the shifting cycle that the Olive villagers still followed on the main island.

4.5 Horticultural Practices and Soil Fertility

The balanced cycle of cultivation and fallow periods has historically played important roles in soil enrichment and food production in the practice of shifting cultivation in Southeast Asia and the islands of the Pacific (Nakano 1978). Table 4.3 shows the reported length of the fallow and cultivation periods for all gardens that the

Fig. 4.7 A man canoeing to a garden on the barrier islands near Raqata in 2001

Table 4.3 Reported length (years) of fallow and cultivation periods and the proportion of gardens made by clearing primary forest, based on location

Locations	Gardens on primary forest [a] /total number of gardens	No. of years: average [range] (*N*) Fallow period	Cultivation period (abandoned [b])	Cultivation period (current [c])
Barrier island	0/13	9.3 [1–40] (12)	29.3 [11–60] (3)	5.0 [0.1–20] (10)
Main island				
Riverside	2/7	27.5 [5–50] (5)	2.8 [1.5–4] (2)	2.9 [1.5–5.1] (5)
Settlement	1/11	15.0 [7–28] (10)	NA [d]	0.65 [0.08–1.33] (11)
Logged lands	23/23	NA (0)	1.7 [0.8–4.2] (17)	1.9 [0.2–4.8] (6)

[a]Primary forest includes forests disturbed by selective logging operations
[b]Gardens that had been abandoned a year before the interview
[c]Gardens that were used during the interview period. Because many villagers had stopped using the barrier island at the outset of the reforestation boom in 2002, the maximum length of the cultivation period for the current gardens was much shorter than that for the abandoned gardens
[d]Because the land in the settlement had not been cultivated until 2002, the data for the abandoned gardens were not available (see Table 4.2)

participating villagers used during the interview period and those they had abandoned up to a year before the interviews, in addition to land uses (primary forest or otherwise) before land was cleared for cultivation.[4] All 13 of the 13 gardens on the barrier island and almost all of the riverside (five of seven) and settlement gardens (10 of 11) were cultivated in areas where they were previously cleared as gardens. Because the villagers had not developed inland lands for cultivation until after the logging road was constructed, they had made all gardens by clearing disturbed primary forest. The limited number of samples in each garden location caused the length of fallow period here to vary widely. The average number of years for the fallow period was 9.3, 15, and 27.5 years on the barrier island, in the settlement and in the riverside gardens on the main island, respectively; in the former, 5 out of 12 gardens laid fallow for three or fewer years. In contrast, the cultivation period was much longer on the barrier island (mean = 29.3 for abandoned gardens and 5.0 for current gardens) than in riverside gardens (2.8 and 2.9 years) and logged lands (1.7 and 1.9 years), respectively. Cultivation of settlement gardens, where data was only available for current gardens, averaged less than 1 year. The limited number of samples, e.g., three for abandoned gardens from the barrier island, seemed to bias the average; however, the range, especially the maximum number of years of cultivation, showed a long cultivation period. Wilcoxon's nonparametric rank sum tests that compared the length of the cultivation period between the barrier and main islands disclosed a significant difference between the abandoned gardens ($W = 63.0$, $P = 0.0072$) as well as a marginally significant difference in the fallow period ($W = 133.0$, $P = 0.0597$).

These findings suggest that shifting cultivation cycles varied between the barrier and main islands. In the latter, the cycle was based on the short cultivation period and long (more than a decade) fallow period, which were generally considered to constitute a sustainable shifting system in a tropical rain forest (Whitmore 1998). However, in the former, a long cultivation period and short fallow period were observed, and this pattern suggested that the cultivated lands seldom shifted and the system will generally lead to an overexploitation of soil nutrients.[5] Another finding was that the gardens on the barrier island, riverside, and settlement were cultivated

[4] For each of the participant households in 2003 (see footnotes 5 and 6), the head and his/her spouse were asked to report their shifting cycles for the gardens that they currently used (hereafter, "current garden") and those they had abandoned during the year before the interview ("abandoned garden"). Limiting these questions to those gardens abandoned within the last year minimized any bias introduced by problems the participants may have had in remembering their past activities. For each garden, the head and spouse of each household were asked if anyone had used the place for horticulture before them (Q1), how long the place had remained unused (or fallowed) before they began cultivation (Q2), and how long they had used (cultivated) the place before they abandoned it or how long they had used (cultivated) the place until the time of the interview (Q3). They were asked to refer to known events such as the end of World War II, the independence of Solomon Islands, the change of the chieftaincy, and the outset of the logging operation if they were unable to report the exact number of years. Their answers were crosschecked using aerial photographs.

[5] However, that type of cultivation was also observed in atoll environments in Melanesia. (Bayliss-Smith 1974).

in a secondary forest, i.e., the villagers reused the lands that had been used previously.

Table 4.4 lists the people who had used the lands before the participants, categorized by their different geographical settings.[6] On the barrier island, 7 out of 13 (54 %) gardens were cultivated in areas that the person himself/herself or his/her parents or siblings had used previously. However, four (31 %) were cultivated in areas that strangers had used. Furthermore, on the main island, both riverside and settlement gardens were cultivated exclusively in the areas that interviewees, his/her relatives, or individuals of the same village had cleared. Fisher's exact test disclosed a significant association between the island and the proportion of gardens in lands once cleared by a non-relative ($P=0.0349$). The villagers claimed that they could cultivate gardens anywhere in Raqata on the barrier island, although permissions were required from their relatives, the elder of the sub-clans, or the paramount chief if any of the villagers wanted to use the main island. According to the villagers, the lands on the riverside and those near settlements on New Georgia Island were divided into small sections, and each sub-clan of the Saikile clan enjoyed privileges in regard to the use of each piece of land.

Table 4.5 shows the nutrient conditions of soils in different geographical settings.[7] The pH and nitrate nitrogen (NO_3-N) levels did not differ significantly between different islands or locations (i.e., barrier island, riverside, settlement, and logged land). However, the soluble phosphoric acid (P_2O_5) level was higher in barrier island gardens (seven out of eight samples were classified at >25 kg P_2O_5/10 ha)

Table 4.4 Users of garden land prior to use by the current villagers, based on location[a]

	Relationship with the participants: *N* (%)		
	Parents, siblings, or self	Individual from the same village	Non-relative in other villages
Barrier island	7 (54)	2 (15)	4 (31)
Main island			
Riverside	3 (60)	2 (40)	0 (0)
Settlement	7 (70)	3 (30)	0 (0)

[a]This table does not show gardens that were made by clearing primary forest (see Table 4.2)

[6] If the head and spouse of each household replied that anyone had used the current garden location for horticulture before them, they were also asked to explain their relationship with that person. The persons using the gardens before the sampled individuals were classified into one of three groups as follows: (1) the participants themselves, or their parents or siblings, (2) someone who lived in the same village as the participants or were genetically related to the participants within two generations, and (3) individuals from another village, having no kinship with the participants for more than at least two generations. Moreover, open-ended interviews were conducted for each participant to reveal the change in land use at each location in the last three to four decades.

[7] In 2003, soil nutrients were measured using a Midori-kun field test-kit (Tokyo University of Agriculture, Tokyo, Japan), which uses a semiquantitative filtration paper to measure the components of soils at a depth of 10 cm. The soil analyses were performed for soils within 3 days of collection.

Table 4.5 A comparison of soil characteristics at barrier and main island locations, including riverside, settlements, and logged land

Items measured	Barrier island	Main island Riverside	Settlement	Logged land	Total	*P*-value: Fisher's exact test (between islands)
N	8	2	9	4	15	
pH						
5.0	3	0	2	1	3	N.S.
5.5	1	2	3	2	7	
6.0	2	0	3	1	4	
6.5	2	0	1	0	1	
NO_3-N (kg/10 ha)						
0	3	0	2	0	2	N.S.
5	0	1	2	1	4	
10	2	0	0	0	0	
15	1	1	2	2	5	
>30	2	0	3	1	4	
P_2O_5 (kg/10 ha)						
<5	1	0	1	0	1	<0.001
10	0	2	6	4		
>25	7	0	2	0	2	

than other lands (only 2 out of 15 samples from main island were at > 25 kg P_2O_5/10 ha) (Fisher's exact test, $P < 0.0001$). The table does not show hydrosoluble potassium (K_2O) because its value was lower than the lowest limit of detection (<5 kg/10 ha) in all samples except for one from the riverside. Neither cultivation nor fallow periods were statistically correlated with any of the soil nutrients. These findings suggest that phosphorus availability was high in the barrier island, despite the short fallow period and the long cultivation period on the island (Table 4.3).

Phosphorus, a primary macronutrient, is a major element in energy transport and transformation processes (e.g., adenosine triphosphate), is a component of phospholipids in plants, and strongly influences root development (Inubushi and Anzai 2001). Potassium is involved in regulating membrane permeability, transpiration, activating enzyme systems, and carbohydrate and protein synthesis. Nitrogen is a major component of proteins and nucleic acids in plant cells and chloroplasts. All of these are essential for the growth of garden crops. Phosphorus was frequently the most limiting factor for plant growth in this study; crops did not respond to nitrogen until adequate phosphorus was available. Although sweet potato and cassava, the main crops in Roviana gardens, were generally recognized as tolerant to low phosphorus soils (Noda et al. 1996), phosphorus deficiency caused nutritional disorders in plants and decreased production, especially during continuous cropping (Goodbody and Humphreys 1986; Hartemink 2003; Hartemink and Bourke 2000; O'Sullivan et al. 1997; Stephens 1969). That is, the cycle of cultivation on the barrier island did not result in the same level of soil degradation when compared with other locations, and high phosphorus availability enabled the use of the shifting cycle of cultivation.

4.6 Crop Diversity and Production from Annuals and Perennials

Data on all foods harvested from gardens were collected every day for 14 days[8]; the author routinely (5–7 times a day) visited the households throughout the day (08:00–20:00).[9] The energy from the foods cultivated in the shifting gardens (annual crops of sweet potatoes, cassava, taro, yam, other root crops, green leaves, and vegetables) was calculated based on the weight of harvested crops and by referring to food composition tables tabulated for the people of the Pacific region (Dignan et al. 1994; Puwastien et al. 2000). Because this study focused on human survival and nutrition with respect to land production, the edible energy yields that villagers had consumed were measured.[10] That is, the crop yield shown in this study was an underestimate of the total harvest from lands.

The measured energy produced per hectare over 14 days was as follows: 2.1, 3.5, 1.0, and 5.3 GJ/ha on the barrier island, on the riverside, in the settlement, and on the logged lands, respectively, indicating high productivity per unit of area on the main island (Table 4.6). This table also shows that the energy produced per unit of time (labor) was highest on the barrier island (5.8 MJ/h), followed by 5.4,

Table 4.6 Total 14-day edible energy yields (MJ per household) of root and other annual crops (green leaves and vegetables), garden area (m^2 per household), time spent for horticulture (hours per household per week), and productivity (edible energy yield/area, edible energy yield/hour), based on location[a]

	Edible energy yield (MJ)	Area (m^2)	Labor time (h)[b]	Edible energy yield/area (GJ/ha)	Edible energy yield/time (MJ/h)
Barrier island	61.5	292.8	5.3	2.10	5.8
Main island					
Riverside	45.7	129.9	4.3	3.52	5.4
Settlement	29.4	290.8	4.7	1.01	3.1
Logged	70.8	133.3	7.4	5.31	4.8
Total	207.3	846.8	21.7	2.45	4.8

[a]The value per household has been shown for convenience, although the analyses were made based on locations

[b]Time spent in transport between the settlement and gardens was also included

[8] In August to September 2003

[9] If foods had been consumed in the gardens or given to other households before the author visited, the participants were asked to report the amount, by referring to the various sizes of samples that had already been measured; these cases accounted for only seven out of 147 recorded household-harvest events.

[10] This estimation was made in tandem with food composition tables (Dignan et al. 1994; Puwastien et al. 2000); edible portions were calculated based on field measurements or previous studies (Kagawa 2001; Umezaki et al. 2001).

4.8, and 3.1 MJ/h for riverside, logged land, and settlement areas, respectively, even though the time for transportation between the settlement and the distant barrier island was included. Obviously, gardening had the best energy efficiency per labor input on the barrier island. The nuts from the local perennial species (i.e., *Canarium indicum* and *Canarium salomonense*) were collected in larger quantities from the barrier island (31.0 MJ per household per week) than from the main island (3.4 MJ).

Table 4.7 shows crops planted in 2003.[11] The plant name and previous botanical research allowed classification of the crops into two types (Hancock and Henderson 1988; Henderson and Hancock 1988): (1) local species that were endemic to the Solomon Islands, neighboring Melanesia, and the Southeast Asian tropics and (2) species that had become naturalized within the last 200 years, i.e., since the first contact with European explorers, or that had been introduced in the last 70 years and did not have local names. Root crops, including both local (e.g., taro and yam) and introduced (e.g., sweet potato and cassava) species, constituted the main diet in all locations. Plants producing edible green leaves, both local and introduced, were also planted on both islands. However, while villagers planted local perennial plants on both islands, the introduced plants, especially those without local names, were rarely planted on the barrier island. For example, two representative local fruit species, malay apple (*hipala* in Roviana vernacular, name: *Eugenia malaccensis*) and papaya (*manioko*: *Carica papaya*), were planted in 20 % and 10 % of the gardens on the barrier island ($N=8$) and 9.1 % and 27.2 % on the main island ($N=25$), respectively; however, introduced jackfruit (*zakifurut* or *beta*: *Artocarpus heterophyllus*), guava (*kuava*: *Psidium guajava*), and pouteria (*poitera*: *Pouteria* spp.) were planted in 31.8 %, 22.7 %, and 18.2 % of gardens, respectively, on the main island, but never on the barrier island. The culturally and spiritually important nut species, *C. indicum* and *C. solomonense*, which are locally called *okete* and *tovinia*, respectively (Hviding 1996; Bayliss-Smith et al. 2003), were observed on both islands. The introduced timber species were never observed on Raqata.

4.7 Carrying Capacity of the Barrier Island

Based on these findings, the land area necessary to sustain the energy needs of Olive villagers and all the other members of the Saikile clan on the barrier island and the main island was calculated for each area. All locations in the main island were pooled because the variations among the riverside, settlements, and logged lands

[11] The head of each household and his/her spouse were asked to report the local names of the plant species planted in each garden, and the author confirmed the existence of the reported plants in the gardens. All plants in the gardens were identified by Roviana vernacular names, and specimens of plants were later identified at the Poitete Forestry School (Mr. Myknee Qusa) and Forestry Station, Munda (Mr. Basile Gua), Ministry of Forest, Environment, and Conservation of Solomon Islands.

Table 4.7 List of species planted in gardens with the proportions (%) of gardens where each species was observed[a]

Common names	Scientific names	Roviana vernacular names	Originality status [b]	Proportion (%) of gardens planted: Barrier island ($N=10$)	Riverside ($N=5$)	Settlement ($N=11$)	Logged land ($N=6$)	Main island total ($N=22$)
Root crops								
Sweet potato	*Ipomoea batatas* L.	*Luzu vaka*	n	100	60	82	83	77
Cassava	*Manihot esculenta* Crantz	*Ore marihi*	n	80	100	100	100	100
Taro	*Colocasia esculenta* L.	*Talo*	e	40	40	64	33	50
Giant taro	*Alocasia macrorrhiza* Schott	*Karuvera*	p	40	0	18	17	14
Yams	*Dioscorea* spp.	*Marihi*	e	40	80	18	33	36
Pana	*Dioscorea esculenta* (Lour.) Burkill	*Dikidiki*	e	30	0	9	33	14
Green leafs								
Slippery cabbage	*Hibiscus manihot* L.	*Neka*	e	70	60	82	67	73
Cabbage	*Brassica chinensis* L.	*Kabisi**	i	40	40	55	33	45
Leeks	*Allium porrum* L.	*Likisi**	i	40	40	45	33	41
Shallot	*Allium ascalonicum* L.	*Saloti**	i	10	0	18	0	9
Polyscias	*Polyscias* spp.	*Tataqala*	e	0	0	9	0	5
Garlic	*Allium sativum* L.	*Gariki**	i	0	0	9	0	5
Water spinach/ Kangkong	*Ipomoea aquatica* Forsskal.	*Kankon**	i	0	0	9	0	5
Other vegetables								
Banana	*Musa* spp.	*Hakua*	e	50	60	45	33	45
Pineapple	*Ananas comosus* (L.) Merr.	*Paenapolo**	i	50	20	27	17	23
Eggplant	*Solanum melongena* L.	*Egipalant**	i	10	0	36	33	27
Chili	*Capsicum annuum* L.	*Sili**	i	10	20	9	33	18
Sugarcane	*Saccharum officinarum* L.	*Suti*	e	10	20	9	17	14

(continued)

Table 4.7 (continued)

Common names	Scientific names	Roviana vernacular names	Originality status [b]	Proportion (%) of gardens planted				
				Barrier island (N=10)	Riverside (N=5)	Settlement (N=11)	Logged land (N=6)	Main island total (N=22)
Ginger	*Zingiber officinale* Roscoe	*Minila*	n	10	20	0	0	5
Peanut	*Arachis hypogaea* L.	*Pinati**	i	10	0	0	0	0
Corn	*Zea mays* L.	*Keto*	i	10	0	0	0	0
Watermelon	*Citrullus lanatus* (Thunb.) Matsum. & Nakai	*Deri*	i	10	0	0	0	0
Common bean	*Phaseolus vulgaris* L.	*Binisi**	i	0	20	36	17	27
Tomato	*Solanum lycopersicum* L.	*Tamata**	i	0	20	27	0	18
Snake gourd	*Trichosanthes cucumerina* L.	*Binisi noki**	i	0	0	27	27	18
Bell pepper	*Capsicum annuum* L. "*grossum*"	*Pepa**	i	0	0	9	33	14
Rice	*Oryza sativa* L.	*Raesi**	i	0	20	9	0	9
Bottle gourd	*Lagenaria* sp.	*Binisi maleisia**	i	0	0	9	0	5
Alpinia pulchra	*Alpinia pulchra* K. Schum.	*Kuruvete*	p	0	20	0	0	5
Fruit tree								
Malay apple	*Syzygium malaccense* (L.) Merr. & L. M. Perry	*Hipala*	e	20	40	0	0	9
Papaya	*Carica papaya* L.	*Manioko*	e	10	0	36	33	27
Tahitian apple	*Spondias dulcis* L.	*Opiti*	e	10	0	0	0	0
Jackfruit	*Artocarpus heterophyllus* Lam.	*Zakifurut/Beta**	i	0	60	9	50	32
Guava	*Psidium guajava* L.	*Kuava**	n	0	40	18	17	23
Pouteria	*Pouteria* spp.	*Poitera**	i	0	40	18	0	18
Soursop	*Annona muricata* L.	*Sasopo/Apolo**	i	0	40	0	0	9
Sugar apple	*Annona squamosa* L.	*Apolo vaka**	i	0	40	0	0	9

Pomelo	*Citrus grandis* L.	*Pomolo**	i	0	20	0	0	5
Granadilla	*Passiflora quadrangularis* L.	*Garandila**	i	0	20	0	0	5
Rambutan	*Nephelium lappaceum* L.	*Ranbutan**	i	0	0	9	9	5
Mandarin	*Citrus reticulate* Blanco	*Mandalin**	i	0	0	0	17	4
Nut tree [c]								
Java almond	*Canarium indicum* L.	*Okete*	e	20	80	9	17	27
Indian almond	*Terminalia catappa* L.	*Tatalise*	e	10	40	9	17	18
Ngali nut	*Canarium salomonense* B. L. Burtt	*Tovinia*	p	0	0	0	17	5
Cutnut	*Barringtonia procera* (Miers) R. Knuth, *B. edulis* Seem.	*Kinu*	e	0	60	27	0	27
Miscellaneous								
Betel tree	*Areca catechu* L.	*Heta*	e	10	60	18	33	32
Betel leaf	*Piper betle* L.	*Igisi*	e	0	40	18	33	27
Timber and cash crops								
Rainbow Eucalyptus	*Eucalyptus deglupta* Blume	*Iuki**	i	0	0	9	50	18
Teak	*Tectona grandis* L.	*Tiki**	i	0	20	18	17	18
Cacao	*Theobroma cacao* L.	*Kokoa**	i	0	20	0	0	5

Asterisks indicate plants without local names

[a]Only plants which were recognized as food or cash crops were included; species used for decoration or other purposes were excluded

[b]The location of the origin of each species is as follows:

p: endemic to New Guinea and Solomon Islands

e: endemic not only in Solomon Islands and Papua New Guinea, but over a much wider area of the tropics

n: species naturalized in the Solomons within the last 200 years, i.e., after the first contact with European explorers; many have local names

i: species introduced in the Solomons within the last 70 years; they usually do not have local names

[c] Nut species present within the garden areas were included; on the barrier island, a number of nut trees grew nearby, but were planted outside gardens

were minor and the number of samples limited. The total arable area (T) necessary for producing the energy necessary to feed K people was calculated as follows: $T = (K \times A \times (R + Y))/Y$ (Carneiro 1964; Rappaport 1984), where K is the total population size, A is the area of cultivated land required to provide the average individual with the amount of food he/she ordinarily derives from cultivated plants per year (ha), and R and Y are the average length of the fallow and cultivation periods (years). The energy required for each of the 112 individuals from the 15 participant households was calculated based on body weight, age, and gender using standards of the Food and Agriculture Organization, World Health Organization, and United Nations University (2004), and their average value (9.3 MJ per person per day) was used as the energy required for the average individual in the Saikile clan. The calculated energy requirement in conjunction with the observed productivity for both the barrier and main islands (Table 4.6) was used to estimate values of A; note that the energy yield of the perennial crops was not included in either garden production or food consumption.

The Saikile clan was estimated to have 2134 members.[12] However, only a portion of them—the actual percentage was estimated to be 90.3 % based on Aswani (2002)—had rights to use the Saikile customary land, including Raqata of Ndora Island.[13] Furthermore, in Saikile, in addition to the neighboring Kalikoqu customary land, the villagers used a territory enclosed entitlement regime model. In this model, land entitlement holders were spatially grouped, territorial boundaries were delineated, and lands controlled by a centralized chieftain authority were recognized regionally; people from outside the community were usually not allowed to use these lands (Aswani 1999, 2002). Based on these conditions, 1934 people were affiliated to the Saikile clan, and the number of people who had rights to use Raqata was estimated to be roughly 2000. Hence, in this study, K was set at 2000 in the scenario estimating the maximum potential land requirement.

Based on the assumption that the energy requirements of all 2000 individuals depended on annual crops, 139.5 ha of the barrier island or 901.5 ha of the main island was required to feed the current population. However, the villagers also used other sources of energy. A previous nutritional study suggested that adult males and females received 39.0 % and 46.0 % of their average energy intake from garden crops, respectively; a second scenario where 42.5 % of energy was obtained from the garden crops suggested that 59.3 ha of the barrier island or 383.1 ha of the main island were required.

[12] As of 2003, based on the latest household census conducted in 1996, when 1,712 members were counted (Statistics Office 1997), and based on the annual rate of increase of 3.2 % for the population of Western Province between 1986 and 1999 (The Solomon Islands Government 2000).

[13] The proportion of households with tribal affiliations to Saikile accounted for 100 %, 98 %, 84 %, and 79 % in the people in the villages of Olive, Nusa Hope, Baraulu, and Ha'apai, respectively (Aswani 2002).

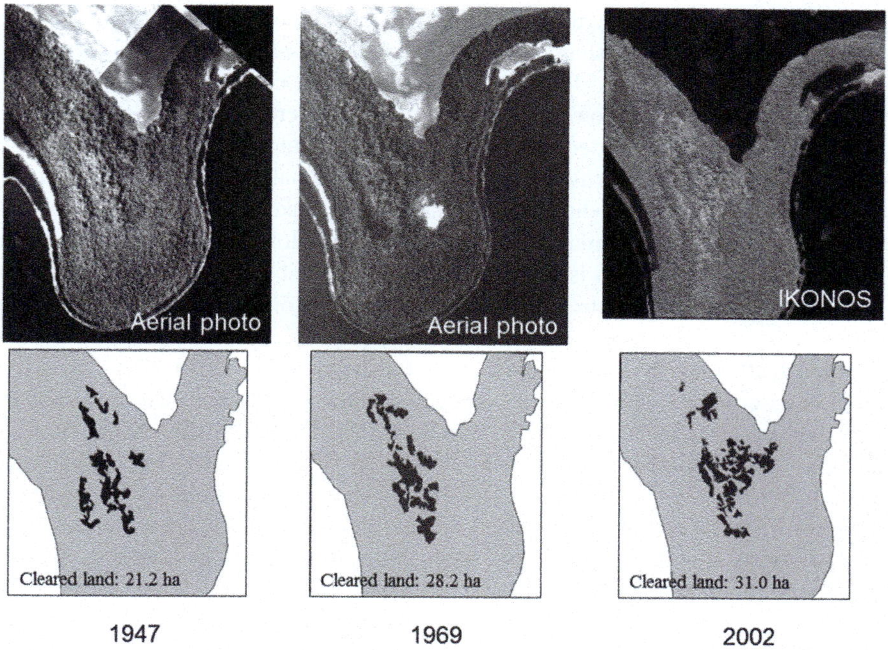

Fig. 4.8 On-screen visual interpretation of cleared land in and around Raqata on the barrier island of Ndora: aerial photographs were taken in 1947 and 1969 and IKONOS satellite image acquired in 2002

To interpret this capacity, aerial photographs that were acquired showing the area in October 1947 and April 1969 and very-high-resolution satellite (IKONOS) data acquired in July 2002 were analyzed. The results suggested that 58.0 ha had previously been cleared in Raqata on the barrier island between 1947 and 2002 (Furusawa et al. 2002); 21.2 ha was cleared in 1947, 28.2 ha in 1969, and 31.0 ha in 2002, but the majority of land was continuously used between these periods (Fig. 4.8).

For comparison, in 1931, the British Solomon Islands Protectorate Government recorded a total number of only 205 people residing in the present Saikile customary land (Jackson 1978). The population in 1996 was 1712, as shown above, indicating the population grew more than eight-fold over 65 years. That is, the increase of cleared land in Raqata was far outpaced by the increase in the human population. Previous studies have suggested that the population had been decreasing in the early twentieth century as a result of the introduction of infectious diseases in this region, so that the population in 1931 might not represent the low point and thus population increase was more rapid than indicated in the present author's analysis (Bayliss-

Smith 2006). Importantly, the lands that had once been cleared (58.0 ha) were almost equal to the lands necessary to feed all Saikile members (59.3 ha). In reality, however, the need for land for subsistence production had not increased dramatically because the production needs from subsistence were decreasing, for example, energy intake from imported foods (e.g., rice) accounted for about 47 % of the energy intake in 2007 (Furusawa et al. 2008). Additionally, substantial amounts of foods had also been produced from the main island. No comparable aerial-photographic data was available for the main island. However, the most important implication of these results is that the Saikile clan was able to sustain its population using the lands that had already been cleared instead of expanding its cultivation areas in the communal land of the barrier island.

In fact, all villagers knew the meaning of the dichotomous terms *toba* (vegetation in the barrier island) and *tutupeka* (vegetation in the main island), and several participants even insisted that the soil in *toba* was more suitable for subsistence production but not for timber species when compared with *tutupeka*. The author also observed that Nusa Hope villagers (approximately 600 inhabitants), who lived and used the Western end of Ndora Island for horticulture, usually canoed to the main New Georgia Island to plant timber species; they stated that such species did not grow well in the *toba* environment.

Overall, the present multidisciplinary research suggested that the productivity per shifting cycle of the barrier island was the highest among the different geographical locations analyzed here. This cycle was supported by natural characteristics (soil rich in phosphorus) and social system (communal use of Raqata). Additionally, the main island was also useful for the villagers for conducting new economic activities such as planting of perennial cash crops rather than for subsistence production. Planting such cash crops was also risky for the villagers, primarily because the trees would cover and occupy the land for a long time and, secondarily, because it was not clear whether such crops would provide them with enough cash income to serve as a substitute for gardening in the future. Thus, the different uses of the barrier and main islands allowed the villagers to integrate new cash crops without the concerns of food security, while receiving the benefits of sustainable production from the barrier island.

4.8 Fishing Activities in the Lagoon and Outer Seas

Maritime anthropological research can also be considered human ecology as it relates to salt water; this type of research has already been intensively studied elsewhere (Aswani 1997). Aswani classified fishing methods into 14 major classes as applied for different types of game and in different ecological settings: netting (*vaqara*), paddle-canoe trolling (*karukarumae*), open sea troll (*karumae makasi*), drift line (*tatadara*), vertical trolling (*kura niugini*), bottom lining (*dakudaku*/*goregore*), sink line (*lolodu*) (Fig. 4.9), angling (*vekovekoe*/*habu tutusa*),

Fig. 4.9 Fishing in a lagoon near Saikile point in 2011

diving (*suvu*), spearing (*hopehopere*), piscicides (*bunabuna*), fish drives (*kuarao*), collecting crabs (*hata garumu/hata kapehe*, etc), and collecting shells (*hata riki/hata deo/hata baŋaopdu*). For example, angling (*habu tutusa*) is done near corals in lagoon water; small fishes caught here were frequently used as bait to catch larger fish in the outer ocean. Trolling (*karu mae*) is done in the outer ocean to catch bonitos and large fish. Net fishing (*vaqara*) was usually done by a group of people in shallow water on a sandy seabed; this kind of zone occurred more frequently in the barrier islands than in areas near the main island. Spearing (*hopohopere*) was done to catch sea turtles in the outer ocean and small bait fishes in the lagoon. Additionally, saltwater habitats were used for the collection of shells and sea cucumber (see Aswani (1997) and Aswani and Lauer (2006) for more details related to the diversity of fishing methods and knowledge).

Use of a portable GPS receiver aided the exploration of how fishing depends on ecological diversity, combined with reference to a number of previous studies (Furusawa 2012; Maddison and Mhurchu 2009; Umezaki et al. 2010). This method allowed the analysis of the intensity of physical activity, broken down by various fishing types and grounds (Furusawa 2012).[14] Voluntary participants were asked to carry GPS units when they left the settlement and conducted subsistence

[14] This research was conducted during a one-week period in July 2007, and follow-up surveys were completed in the following weeks. See Furusawa (2011) for information on the technical limitations of adopting this method for the convenience of future fieldworkers.

Table 4.8 Participant identification and recorded fishing and marine resource gathering activities

Participant ID	Gender	Age	Reported activities	Main activity places
A	M	30	Angling	Inner lagoon reef
B	M	31	Angling; trolling; firewood collection	Inner lagoon reef
C	M	27	Trolling; angling	Inner lagoon reef; outer reef drop; passage
D	M	31	Trolling	Outer reef drop; open ocean
E	M	37	Trolling; angling	Open ocean; passage
F	M	38	Trolling	Open ocean; passage
G	M	52	Collection of mud clams/nassa shells	Shallow sandy and reefs
H	F	43	Collection of mud clams/nassa shells	Shallow sandy and reefs
I	M	35	Collection of mud clams	Shallow sandy and reefs
J	M	59	Collection of nassa shells; angling	Shallow sandy and reefs
K	M	28	Collection of sea cucumber	Shallow sandy and reefs
L	M	21	Collection of oyster	Mangrove

activities.[15] Fifty activities of 38 villagers were thus recorded; of those, 12 activities of 12 different villagers were related to fishing and marine resource collection, respectively (Table 4.8). Each participant was also interviewed about the exact location and activity of the day when they returned.

Participants A through F mainly engaged in fishing, while G through L participated in the collection of shells. Males engaged in angling, trolling, and other kinds of fishing, while females participated in collecting marine resources (Table 4.8). The Olive villagers' fishing activities included foraging in the inner lagoon reefs (*sagauru* in local Roviana language), outer barrier reef edges where the reefs dropped off into deep water (*vuragare*), lagoon passages (*holapana*), and open ocean (*kolo lamana*) and also included transport by canoe (*vose*) between the settlement and these fishing points. The GPS tracks were thus classified into canoeing in

[15] The GPS unit was packed in a small (15 cm × 16 cm) water-proof PET electronic protection bag (Seal Line; Cascade Designs, Inc., Seattle, WA, USA). Each participant wore this bag on his/her waist. The Foretrex 101 (Garmin, Ltd., Olathe, KS, USA) is a battery-driven, water-resistant, and wristwatch-size GPS unit that weighs 78 g. An internal memory card provides the unit with the capacity to store 10,000 points, and two alkaline AAA batteries can run the system for approximately 15 hours. The units were set to record the GPS coordinates of their locations at five-second intervals with the Wide Area Augmentation System enabled. Rodriguez et al. (2005), in their study of residents' daily lives in a town in North Carolina, USA, confirmed that a wristwatch-size GPS unit (Foretrex 201, Garmin Ltd.) had a high level of accuracy; the average distance from the recorded points to the geodetic point was 3.02 m, while 81.1 % and 99.9 % of points were located within 5-m and 20-m buffers, respectively.

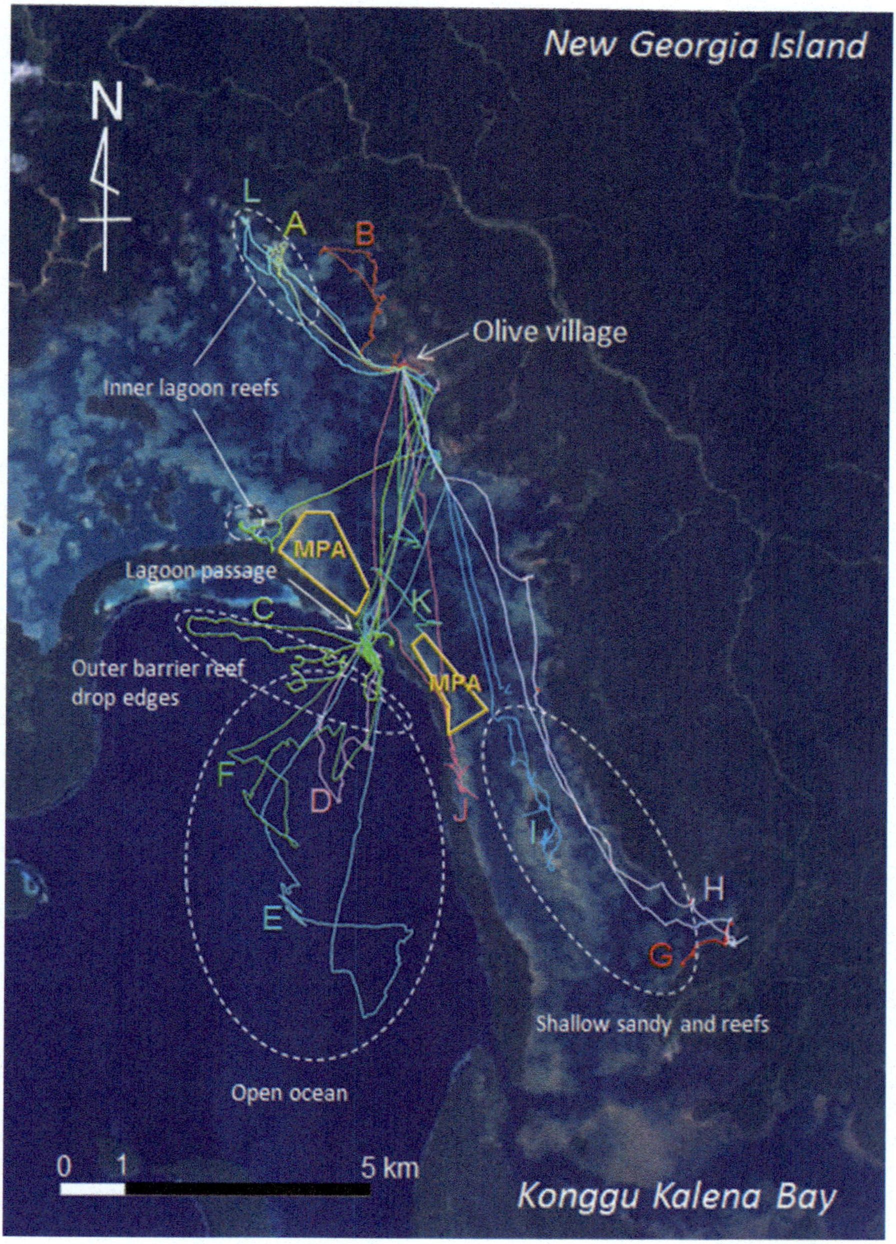

Fig. 4.10 Twelve fishing trips (A to L) tracked by a portable Global Positioning System

Fig. 4.11 *Nassarius* shell collected and sold to a store at Olive in 2007

the inner lagoon; canoeing in the open ocean; and fishing, including both angling and trolling, in the inner lagoon, passages, outer reef drop-off edges, and the open ocean. Detailed analyses were made only for fishing because the data acquisition was poor in regard to the shell collection activities. Using a geographic information system (GIS), track data were overlain on a layer of a Landsat ETM+ image (acquired on December 1, 2002; NASA Landsat Program, USGS, Sioux Falls, USA) along with digital elevation data (Shuttle Radar Topography Mission) acquired from the Global Land Cover Facility of the University of Maryland, USA; these layers were reviewed and interpreted individually.[16]

Figure 4.10 shows the locations of the 12 recorded trips. The villagers angled (*habu tutusa*) in the inner coral reefs and a lagoon passage. They trolled (*karu mae*) in the open ocean to catch bonitos, in the inner lagoon mainly for trevallies, and in the outer barrier reef edges for various types of large fish (e.g., barracudas). Trolling in the open ocean used a wide area of ocean. The villagers reported that the *Nassarius* shells occur in abundance in the sandy shallow inner lagoon of the distant areas of Koqu Kalena Bay (Fig. 4.11), while sea cucumbers are available in the shallow sea closer to the settlement and shells for consumption were found in mangroves. Additionally, the villagers frequently combined two or more activities in one trip. For example, participant B went fishing in an inner lagoon and then visited his garden and forest for firewood collection during the same trip. Participant D trolled in

[16] The GPS record provided the location and the rate of movement during a trip. Because the author had identified the locations and activities that the villagers used in fishing during previous fieldwork, the precise locations and activities based on the GPS record in tandem with the participants' detailed reports were identified.

the open ocean and then angled and trolled in the outer barrier reef edges. This figure clearly showed the villagers used a wide area of sea and a variety of marine ecosystems by traveling more than 30 km before returning home.

In summary, the villagers chose appropriate lands for both cultivation and activities that generated a cash income, as well as for the use of marine resources, employing both efficient production methods and using sustainable production techniques. Additionally, the villagers' subsistence lifestyle depended on the diversity of the landscape and the availability of various species.

Chapter 5
Plant Resources as Ecosystems Services

Abstract This chapter employs a quantitative ethnobotanical analysis to explore the kinds of plants the people use for their daily life and compares the local plant and land use patterns of a rural village and an urbanized village. This chapter then addresses how local people depend on biodiversity and how traditional human modifications of the forest contribute to the conservation of biodiversity. As a result, the proportion of useful trees was high in the barrier island's primary forest (68.4 %) and the main island's reserve (68.3 %) and low in the main island's secondary forest (29.3 %). Some species were found only in human-modified forests (e.g., reserve and secondary forests). Additionally, the reserve and mangrove forests, which existed exclusively in the rural village area, provided the rural people with a number of useful resources, while the urban people lacked access to these resources. These findings illustrate how a local society's use of resources was related to community welfare as well as to environmental preservation.

Keywords Plant use • Food • Medicine • Building material • Tool • Ecosystem services • Distribution of useful plants • Subsistence and diversity

5.1 Quantitative Ethnobotany

Every society has also developed indigenous botanical knowledge (hereafter called ethnobotanical knowledge) through its dynamic interactions with the surrounding floral environment (Townsend 2000). This type of knowledge provides a valuable and even indispensable resource not only for the survival of a particular society but also for all of humanity, because ethnobotanical knowledge potentially offers new products and genetic resources to the Western world that can be made available globally. Recently, several researchers have not only recorded and analyzed production and consumption in the Solomon Islands and other Melanesian and Southeast Asian societies in various ways but have also studied their biocultural diversity (Akimichi 1978; Hviding 2005; Kwa'ioloa and Burt 2001; Pikacha 2008; Sillitoe 1998; Takekawa 2000). Many such studies have integrated ethnobotany into their analysis, which is devoted to understanding how local people recognize, use, and

T. Furusawa, *Living with Biodiversity in an Island Ecosystem*, Ethnobiology,
DOI 10.1007/978-981-287-904-2_5

manage plant resources, and have investigated how ethnobiology can be applied to the conservation of biodiversity. First, this can be done by learning and understanding the ethnobotanical knowledge that local peoples have acquired regarding sustainable or minimally destructive methods of using and managing resources (Plotkin 1995) and, second, by revealing the identity of important species that potentially require protection (Prance et al. 1987).

Prance and others (1987) developed "quantitative ethnobotany" to measure the cultural importance of each species used by different societies and in different environments in their study on the Amazon River drainage. Systematic quantification with detailed observation of the behaviors and lifestyles of individual people will allow researchers to improve the measurement of usefulness and importance of different species and different types of land use (Furusawa 2009). This method also identifies important and useful vegetation or land cover types that local people employ.

This chapter explains how people use plants, especially trees (including palms), to obtain a variety of resources. Special attention was paid to compare how rural and urban residents used plant resources differently, i.e., the effects of acculturation on plant use. Additional analyses explore the interaction between these two kinds of "diversity," allowing a discussion of the potential interactions between humans and the natural world. This chapter finally documents how the people of Roviana traditionally use each of several forest types and how each forest type supports different plant species that are used for various purposes.

5.2 Plants Used for Food

The descriptions used here are mainly based on research conducted in the Roviana villages Olive and Dunde in 2003.[1] Other qualitative and quantitative research conducted during a different time period provides supplemental information. Appendix 1 provides a list of plants that people of these villages were observed to use.

Plants used as food serve as our first category. We observed all foods consumed in each study household during a 1-week period and recorded their weights every 80–90 min from 7:00 a.m. to 8:20–8:30 p.m. every day.[2] When a participant ate food outside the village, the diet information was obtained through an interview. The

[1] For detailed observations of plant use, 15 and 17 households were randomly selected from the 65 and 202 households in Olive and Dunde, respectively. Written informed consent for participation in the survey was obtained from each head-of-household (head). The participants were informed that they could withdraw from the study at any time and that they had a right to refuse to answer any questions. All interviews were conducted in the local Roviana language by the author with the help of local assistants (Mr. Edwin Huti and Mr. Rex Daga). No economic incentives were provided to the participant households to avoid bias; however, following completion of the research, a suitable cultural gift (food) was given to each household.

[2] The food surveys were conducted in August in both villages to avoid seasonal differences between the two villages.

energy (MJ) contribution of the plants was estimated using measurements and interviews in tandem with food composition tables (Dignan et al. 1994; Puwastien et al. 2000).

The interviewees identified 39 species used as food. Sweet potato (*I. batatas*) and cassava (*M. esculenta*) were the primary sources of energy in both villages (38.7 % and 18.3 % in the rural and urban villages, respectively; Tables 5.1 and 5.2).[3] The villagers typically ate these root crops simply by boiling or baking them on most days (Fig. 5.1), but they also mixed cassava (traditionally taro or yam) with coconut milk and pounded the mixture well to make a cake called *nuqara* (Fig. 5.2). This *nuqara* was eaten raw but was sometimes baked in a stone oven (*motu*) to make a baked pudding called *inasana ore-marih. Nuqara* and *inasana* were frequently made by mixing them with Java almonds (also known as kenari nuts) (*okete*: *Canarium indicum*), ngali nuts (*tovinia*: *Canarium salomonense*), or banana. *Nuqara* and *inasana* were always prepared for events such as feasts after church services every Sunday or for big events such as birthday feasts of the chiefs. According to informants, traditionally a layered cake (called *mamahi*) made using *nuqara* (with nuts) and slippery cabbage (*neka*: *Hibiscus manihot*) was a special food item that only a chief was allowed to eat.

Table 5.2 shows that imported rice constituted a substantial proportion of an individual's energy intake and served as the main food energy source consumed in Dunde. Additionally, in Dunde, a type of flour and cooking oil were baked to make bread and a type of local doughnut (20.0 % in total) also provided a large amount of energy (also discussed in Chap. 9).

Table 5.1 Food items consisting of more than 2 % of an individual's energy intake in Olive

Food item	Scientific name	Roviana name	% in total energy intake
Cassava	*Manihot esculenta* Crantz	*Ore marihi*	27.9
Rice[a]		*Raesi*	23.5
Sweet potato	*Ipomoea batatas* L.	*Luzu vaka*	10.8
Sugar[a]		*Suga*	9.0
Coconut	*Cocos nucifera* L.	*Ŋohara*	5.5
Canarium nut	*Canarium indicum* L. and *Canarium salomonense* Burtt	*Okete/tovinia*	3.3
Local doughnut[b]		*Girisi*	3.0
Fish cooked with coconut		*Igana raro Ŋohara*	2.1

[a]Imported or industrially processed foods
[b]Local doughnuts are made from flour and cooking oil. These are homemade, but are sometimes sold in the settlement

[3] Although the villagers planted a number of varieties of these root crops, all varieties were grouped at the species level.

Table 5.2 Food items consisting of more than 2 % of an individual's energy intake in Dunde

Food item	Scientific name	Roviana name	% in total energy intake
Rice[a]		*Raesi*	26.1
Sweet potato	*Ipomoea batatas* L.	*Luzu vaka*	13.9
Bread[b]		*Bani, buredi*	10.1
Local doughnut[c]		*Girisi, donati*	9.9
Cassava	*Manihot esculenta* Crantz	*Ore marihi*	4.4
Coconut	*Cocos nucifera* L.	*Ɗohara*	4.3
Fish		*Igana*	3.3
Crab		*Garumu*	3.2
Tinned tuna[a]		*Taiyo*	3.1
Sugar		*Suga*	2.4
Banana	*Musa* spp.	*Hakua*	2.2

[a]Imported or industrially processed foods
[b]Bread is mainly made from flour and is then baked and sold by villagers
[c]Local doughnuts are made from flour and cooking oil. These are homemade, but are sometimes sold in the settlement

Fig. 5.1 Peeling sweet potato and cassava for cooking at Olive in 2005

Fig. 5.2 Pounding cassava with coconut milk to make a *nuqara* cake at Olive in 2005

Table 5.3 Six tree or palm species used as food; these six contributed most of the energy intake from trees and palms

Common name	Scientific name	Roviana name	% in total energy intake	
			Rural	Urban
Coconut	*Cocos nucifera* L.	*Ŋohara*	4.3	3.9
Canarium nut[a]	*Canarium indicum* L. and *Canarium salomonense* Burtt	*Okete/tovinia*	3.9	0.1
Banana	*Musa* spp.	*Hakua*	0.4	2.2
Papaya	*Carica papaya* L.	*Manioko*	0.1	0.6
Large-leafed mangrove	*Bruguiera gymnorrhiza* (L.) Lam.	*Petu*	0.04	0
Gnetum	*Gnetum gnemon* L.	*Leqe*	0	0.02

[a]Java almond and ngali nut were grouped as one here, because the Roviana term *okete* is sometime used as a general term meaning either species of the genus *Canarium*

Six representative tree or palm species (Table 5.3) provided the next highest levels of energy after sweet potato and cassava (also see Chap. 4 for other crops used in Roviana). Coconut contributed about 4 % of the total energy intake in both villages. A Roviana term, *ŋohara*, is used to mean both a coconut tree itself and a matured fruit of coconut, while another term, *bulo*, refers to a young coconut which people use to obtain coconut water. Two species of *Canarium* nuts (Java almond, *C. indicum*; ngali nut, *C. salomonense*) accounted for 3.3 % of the energy intake in the rural area but only 0.1 % of the energy intake in the urban area. Because the Roviana term *okete* means not only the single species of Java almond but is also used for the food nut of all plants in the genus *Canarium*, all *Canarium* species were combined in the data analyses of food intake. While most of the food plants were cultivated or planted in gardens, gnetum tree (*leqe*: *Gnetum gnemon*) grew in *nobo* secondary forest (Chap. 3); while people did not always plant this tree, they usually refrained from weeding or damaging it so that it grew under semidomesticated conditions. Semidomesticated means a plant has not been fully domesticated, but is protected by people when found in the wild and near settlements or gardens (Kobayashi et al. 2013; Matsui 1989). Fruits were collected from wild large-leafed mangrove (*B. gymnorrhiza*).

5.3 Plants Used to Treat Illnesses

This section reports how plants were used to treat illnesses. Each household was visited every evening, and the head and/or spouse were asked to report any illnesses and describe any treatments, including the use of medicinal plants that were used in that household for 28 days.[4] The author tallied the total number of ill person-days and the numbers of ill person-days on which traditional and Western medicines were used (Furusawa 2006). On some days, the ill person used both traditional and Western treatments.

Herbal medicines were used in 14.5 % and 25.4 % of all treatments for ill person-days, accounting for 201 and 159 cases in the rural and urban areas, respectively. Nineteen species were used on 112 person-days in both village areas combined. Table 5.4 shows representative species used for treatment. Coconut oil, the most common herbal treatment, was used on 30 person-days to treat wounds and skin conditions or was also used during massage treatments that were done to relieve pain. Coconut oil mixed with oil extracted from ylang-ylang (*naqarita*: *Cananga odorata*) was used on 25 person-days. Rose periwinkle (*Catharanthus roseus*), which was recently introduced as a flowering and ornamental plant in urban areas and lacked a Roviana name, was used on 16 person-days for diabetes. Heartleaf hempvine (*Luluzu*: *Mikania cordata*) was used to treat wounds (Fig. 5.3). In addition to coconut palm and ylang-ylang, three species, tropical almond, betel nut palm (*A. catechu*), and great morinda (*gurata*: *Morinda citrifolia*), were used for 2.0–2.5 % of all ill person-days in the two villages, respectively. Medicinal species were used more frequently in the rural areas. Almost all herbal plants were planted in the settlements or grown under semidomesticated conditions in secondary growth nearby, such as tropical almond and *Cassia alata* (one person-day in the rural area for the latter).

[4] When the patient consulted other villagers or herbalists outside of the village, the herbalist was interviewed about the recipe. If the herbalist refused to disclose his/her recipe, that treatment was excluded from the analysis.

Table 5.4 Species used to treat illnesses for two or more person-days

Common name	Scientific name	Roviana name	Frequency of use: no. of person-days (%)			
			Rural		Urban	
Coconut	*Cocos nucifera* L.	*Ŋohara*	5	(2.0)	25	(15.7)
Ylang-ylang	*Cananga odorata* (Lam.) Hook. f. & Thoms.	*Naqarita*	0	(0)	25	(15.7)
Rose periwinkle	*Catharanthus roseus* (L.) G. Don	NA	0	(0)	16	(10.1)
Tropical almond	*Terminalia catappa* L.	*Tatalise*	5	(2.0)	0	(0)
Heartleaf Hempvine	*Mikania cordata* (Burm. f.) B. L. Rob.	*Luluzu*	5	(2.0)	1	(0.6)
Betel nut	*Areca catechu* L.	*Heta*	4	(2.0)	0	(0)
Betel leaf	*Piper betle* L.	*Igisi*	4	(2.0)	0	(0)
Great morinda	*Morinda citrifolia* L.	*Gurata*	0	(0)	4	(2.5)
Timonius timon	*Timonius timon* (Spreng.) Merr.	*Toqoŋeta*	2	(1.0)	0	(0)
Red clerodendrum	*Clerodendrum buchananii* (Roxb.) Walp.	*Titimunuhaha*	2	(1.0)	0	(0)
Native mulberry	*Pipturus argenteus* (Forst. f.) Wedd.	*Nekete*	2	(1.0)	0	(0)
Crinum lily	*Crinum* sp.	*Vogi*	2	(1.0)	0	(0)
Key lime	*Citrus aurantifolia* (Christm.) Swing.	*Laini*	1	(0.5)	1	(0.6)

5.4 Plants Used to Build Houses

The author visited every household and asked the head and/or builder to provide the local Roviana name for each material used during construction, i.e., the names of the materials used for floors, walls, poles, rafters, beams, roofs, and other parts of the houses. Seventy-one plant species were used for building materials, including trees, palms (including rattan), and bamboo. Table 5.5 shows the 11 species most frequently used for construction by the households in both villages. The two most frequently used species were *Calophyllum* spp. (Calophyllum or *buni*[5]; 100 % in each village) and *V. cofassus* (*vasara*; 100 and 88.2 % in the rural and urban villages, respectively). These trees were too large for the rural people to fell and cut up by themselves without chainsaws; therefore, the villagers frequently purchased

[5] People commonly use a type of *buni* called *buni masa* (literally meaning "*buni* at the coast"; *Calophyllum inophyllum*); however, the *buni* used for building houses includes various *Calophyllum* species grown in forest.

Fig. 5.3 Treating a wound with a leaf of heartleaf hempvine (*Luluzu*)

sawn timber from the logging company campsites. Local markets sell low-grade timber at low prices. Sago palm (*edeve*: *Metroxylon* spp.) leaves were the major materials used for walls and roofs in traditional leaf houses (Fig. 5.4); the palm leaves were tightened with thin sticks made from local areca nut palm trunks (*heta pinomo*: *Areca macrocalyx*). Rattan palms (*aroso*: *Calamus* spp.) were used as ropes to tighten joints and for many other parts of the houses. Iron nails were rarely used in building traditional leaf houses. Although Western-style permanent houses were built exclusively with sawn timbers (Fig. 5.5), many households living in such houses had small kitchen huts built, in part, with rattan (93.3 % and 82.4 % in the rural and urban villages, respectively). Medium-sized Flueggea trees (*mavuana*: *F. flexuosa*) were a major source of posts used to build leaf houses; therefore, this tree was used more in the rural village (93.3 % of all rural houses) rather than in the urban village (70.6 %). This tree was used in a round log form after removal of its bark. Fijian longan (*qema*: *P. pinnata*) was frequently used in the urban village (94.1 %) and rarely used in the rural one (20.0 %). Loggers harvest this tree species and traded them to villages at the logging campsites in this area; this large tree primarily grows on the barrier islands. Because logging operations have been prohibited on the barrier islands of the Saikile clan until recently, the timbers of this tree were rarely used in the rural village. Large-leafed mangroves (*petu*: *B. gymnorrhiza*), which were more abundant in the rural areas, were more frequently used in the rural village (66.7 %) than in the urban village (47.1 %). Trees that were abundant in

Table 5.5 Eleven tree or palm species used in the highest frequencies for building houses

Common name	Scientific name	Roviana name	Frequency of use: % household using		Main purposes
			Rural (N=15)	Urban (N=17)	
Calophyllum	*Calophyllum* spp.	*Buni*	100	100	Floor, wall, post
Vitex	*Vitex cofassus* Reinw. ex Bl.	*Vasara*	100	88.2	Floor, wall, post
Sago palm	*Metroxylon* spp.	*Edeve*	93.3	82.4	Roof, wall
Rattan	*Calamus* spp.	*Aroso*	93.3	82.4	Rope
Betel nut palm (wild)	*Areca macrocalyx* Zipp. Ex Bl.	*Heta pinomo*	93.3	82.4	Rafter, wall, floor
Flueggea	*Flueggea flexuosa* Müll. Arg.	*Mavuana*	93.3	70.6	Post
Fijian longan, taun	*Pometia pinnata* Forst. f.	*Qema*	20.0	94.1	Floor, wall, post
Large-leafed mangrove	*Bruguiera gymnorrhiza* (L.) Lam.	*Petu*	66.7	47.1	Post, rafter
Brown kurrajong	*Commersonia bartramia* (L.) Merr.	*Zamara*	66.7	0	Rafter, beam
Timonius timon	*Timonius timon* (Spreng.) Merr.	*Toqoneta*	13.3	41.2	Beam
Dillenia salomonensis	*Dillenia salomonensis* (C. T. White) Hoogl.	*Kapuhu*	46.7	0	Floor, post

secondary growth (brown kurrajong (*zamara*: *C. bartramia*) and *D. salomonensis* (*kapuhu*)) were used as materials for building leaf houses and were used exclusively in the rural village.

5.5 Plants Used for Tools

The use of tools made from plants was observed and was also discussed during surveys and interviews. First, the head and/or spouse were asked to list all plant-made tools they owned. A single, village-wide list of tools was compiled by visiting each household in turn. Then, each household was visited again, and the heads/spouses were asked what materials were used for each item on the final list of 27 types of tools. Note that any plants used for rituals and magic were also categorized as tools.

Fig. 5.4 Building a traditional leaf house. The roof is made from sago palm leaves sewn onto the trunks of wild betel nut palm

Fig. 5.5 A Western-style permanent house in Dunde

Among the various daily commodities, 27 tools were made from 53 species of plants. Table 5.6 lists the 13 tree or palm species that were used most frequently for tools. Vitex (*vasara*: *V. cofassus*) was used in all households in both villages to make paddles, wood mortar, agricultural tools, and furniture. Kapok and *B. malabaricum* fiber were used for pillows in almost all households (86.7 % and 100 % in the rural and urban villages, respectively). Coconut palm leaves were used for hats, bags, fans, and other items, while sago palm leaves were used for brooms. White beech (*qoliti*: *G. moluccana*) was one of the most important tree species because canoes are made exclusively from this species (80.0 % and 70.6 % in the rural and urban households owned canoes, respectively); note that a canoe was sometimes used by two or more households (e.g., shared in a kinship group), and thus, all households

Table 5.6 Thirteen tree or palm species used in the highest frequencies as tools

Common name	Scientific name	Roviana name	Frequency of use: % household using		Main purpose
			Rural (N=15)	Urban (N=17)	
Vitex	*Vitex cofassus* Reinw. ex Bl.	*Vasara*	100	100	Paddle, Furniture, Mortar, Plow
Kapok	*Ceiba pentandra* (L.) Gaertn. *Bombax malabaricum* DC.	*Lozi*	86.7	100	Pillow
Coconut palm	*Cocos nucifera* L.	*Dohara*	86.7	100	Basket, Hat
Sago palm	*Metroxylon* spp.	*Edeve*	93.3	82.4	Broom
Rattan	*Calamus* spp.	*Aroso*	86.7	88.2	Chair, Tong
Calophyllum	*Calophyllum* spp.	*Buni*	93.3	76.5	Spear hand, Furniture
Fijian longan, taun	*Pometia pinnata* Forst. f.	*Qema*	80.0	82.4	Axe hand
Pemphis	*Pemphis acidula* J. R. & G. Forst.	*Bobogele*	60.0	94.1	Pestle, Coconut opener
White beech	*Gmelina moluccana* Backer ex K. Heyne	*Qoliti*	80.0	70.6	Canoe
Pandanus	*Pandanus* sp.	*Dalou*	80.0	41.2	Mat
Premna	*Premna corymbosa* Rottler & Willd.	*Zovi*	93.3	23.5	Amulet
Pandanus tectorius	*Pandanus tectorius* Park.	*Pate*	40.0	29.4	Mate
Large-leafed mangrove	*Bruguiera gymnorrhiza* (L.) Lam.	*Petu*	66.7	0	Pestle, Coconut opener

had access to canoes. A significant difference between the two villages (66.7 % in the rural village, 0 % in the urban one) was found in the use of large-leafed mangroves, because mangrove forests are abundant only in the rural village. The mangroves were used as sticks or knives to open coconut fruit (*viviguana* in the Roviana language). A difference was also observed (93.3 % and 23.5 % in the rural and urban villages, respectively) in the use of premna (*zovi*: *Premna corymbosa*) for amulets, including amulets used as good luck charms in fishing or to ward off devil spirits, for example (Fig. 5.6). This customary charm was less frequently found in the urban village.

In the urban village, materials that were not available from forests were purchased. For example, large-leafed mangrove sticks were used to pry open coconuts (*viguvigua* in Roviana) in nine of 15 households in the rural village, while imported iron was used for this purpose in the urban village.

In this study, all canoes observed were dugout canoes (*hore*), but two other types of canoes had been employed in the past called the *mola* and *tomoko*, both of which were constructed by binding different parts to form the canoe; a *tomoko* is a war canoe previously made in this area (Fig. 5.7). According to an elder informant, light trees, such as *Palaquium erythrospermum*, were used to build war canoes, but all canoes today are made from white beech.

5.6 Plants Used for Fuel

By observing all trees collected as firewood over a 14-day period, the names of 44 fuel species were recorded during interviews. According to the village interviewees, however, people used all trees as firewood without being selective. People in the urban village were frequently observed using demolition debris (wood) from houses and broken tools as fuel. Also, coconut husks were frequently used for making and transferring fire. A few people refrained from using local *Canarium* nuts (*C. indicum* or *C. salomonense*) for fuel, which were customarily important nuts, as well as *G. celebica* and *P. corymbosa*, because these species were sacred for them or their clans.

5.7 Plants Used in Two or More Purposes

This chapter examined four categories of use for plants: food, medicine, building materials, and tools. In total, villagers used 149 species for these four uses during the study period; trees used only for fuel (11 species) were excluded from this analysis because all trees were used for fuel with little selectivity. Among them, coconut was represented in all four use categories, suggesting that this plant is the most important species for the Roviana people. Coconut is important in the Pacific Island societies because it provides a high-calorie source of food, building materials,

Fig. 5.6 Premna (*zovi*) leaves carried on a boat as a fishing amulet

Fig. 5.7 *Tomoko*—recently redesigned for ceremonial reasons (*foreground*)—and dugout canoes (*background*). *Duzuŋuzu*, an image of spiritual power, is attached on the *front end* (*left* in photo) of the *tomoko*

herbal medicines, and fuel (Dignan et al. 1994; Henderson and Hancock 1988; Hviding and Bayliss-Smith 2000). A mangrove species (*B. gymnorrhiza*) was used for foods, buildings, and tools. A previous study also found mangroves have various purposes (i.e., food, medicine, building materials, tools, firewood) in Marovo (Hviding and Bayliss-Smith 2000), and mangroves are believed to have traditionally played important roles in the subsistence of the Western Solomon Islanders. *Timonius timon*, which is not commonly used in other societies, was used for medicine, building materials, and tools. This tree grows wild on cleared land and roadsides. The people weed its seedlings out of their gardens or plantation areas but leave it untouched elsewhere; additionally, they sometimes aid the growth of nearby *Timonius* trees by cutting or weeding out trees that might slow the growth to the *Timonius* trees. Such behaviors, called "semidomestication," were observed with other plants, such as gnetum and premna. The villagers thus enjoyed a range of ecological services provided by the biodiversity that was preserved in their forests.

5.8 Distribution of Useful Trees in Different Forest Types and on Islands

Further analyses were made to explore the interaction between these two kinds of "diversity," allowing discussion of the potential interactions between humans and the natural world. Many tree species were observed in the plant-use survey quadrats. These included trees growing in five types of forests: (1) primary, (2) reserve, and (3) secondary forests on the main island as well as (4) primary and (5) secondary forests on the barrier island; a total of 87 individual (15 species) and 82 (26), 53

Table 5.7 Numbers of trees, species, and proportions of useful trees found in each forest type

	No. of trees (per 1/4 ha)	No. of species (per 1/4 ha)	No. of useful trees (per 1/4 ha)	No. of useful species (per 1/4 ha)	% of useful trees	% of useful species
Main island (*Tutupeka*)						
Primary	168	31	87	15	51.8	48.4
Reserve[a]	120 (122, 117)	49	82	26	68.3	53.1
Secondary	181	48	53	18	29.3	37.5
Mangrove	137	10	87	6	63.5	60.0
Logged (selectively)	180	49	92	20	51.1	40.8
Barrier island (*Toba*)						
Primary	117	19	80	12	68.4	63.2
Secondary	124	36	77	16	62.1	44.4

[a]Two quadrats (1/4 ha each) were made and the number of trees were averaged for the two reserve forests analyzed here (the numbers of trees respective quadrats are shown in parentheses)

(18), 80 (12), and 77 (16) trees were found in those habitats, respectively (see Appendix); trees used only for fuels were excluded from this analysis. Additionally, a total of 87 trees (6 species) were also found in the mangrove quadrats (Table 5.7). The proportion of useful trees was high in the barrier island's primary forest (68.4 %) and the main island's reserve (68.3 %) and low in the main island's secondary forest (29.3 %).

5.9 The Relationship between Subsistence Plant Use and Botanical Diversity

Figure 5.8 shows the proportions of plants used for building materials, tools, and fuels in the different forest types by the urban and rural people. This analysis implies that plant collection sites used in this study contain a large amount of biomass and support a variety of tree species used for many purposes, in addition to supplying fuelwood. Trees from primary forests used as building materials accounted for 36.0 % of the tree species used by urban villagers and only 26.6 % of the tree species used by rural villagers ($\chi^2=7.7$, $P=0.004$). However, most of the building materials in the former group were timbers purchased from logging companies operating in neighboring areas. In the reserve forests, 52.3 % of the trees were species used by the rural people as building materials; this proportion was only 18.4 % in the urban village ($\chi^2=58.6$, $P<0.001$). Urban villagers seldom used trees found in the mangrove forests to make tools, probably because few mangroves remained in that village, while 62.3 % of the tree species in the mangroves were used in the rural village. Rural villagers found more tree species in the reserve forests useful (16.7 %) than urban villagers found useful (6.7 %; $\chi^2=10.7$, $P=0.001$). Higher proportions of tree species in the reserve, secondary, and mangrove forests were used for fuels in the rural village than in the urban village. These results suggest that the reserve and mangrove forests, which existed exclusively in the rural village area, provided the rural people with a number of useful resources, while the urban people lacked access to these resources. Chapter 9 explains the economic burdens of the urban people.

5.10 Traditional Methods of Subsistence May Contribute to Forest Biodiversity

Figure 5.9 shows the numbers of trees of representative useful species found in the respective forest types. Calophyllum, a primary building material, was predominantly found on the main island and in primary and reserve forests. Dillenia (*Dillenia salomonensis*) was found only in the reserve forests. Two useful species were found only in the secondary forests: Commersonia (*Commersonia bartramia*), an

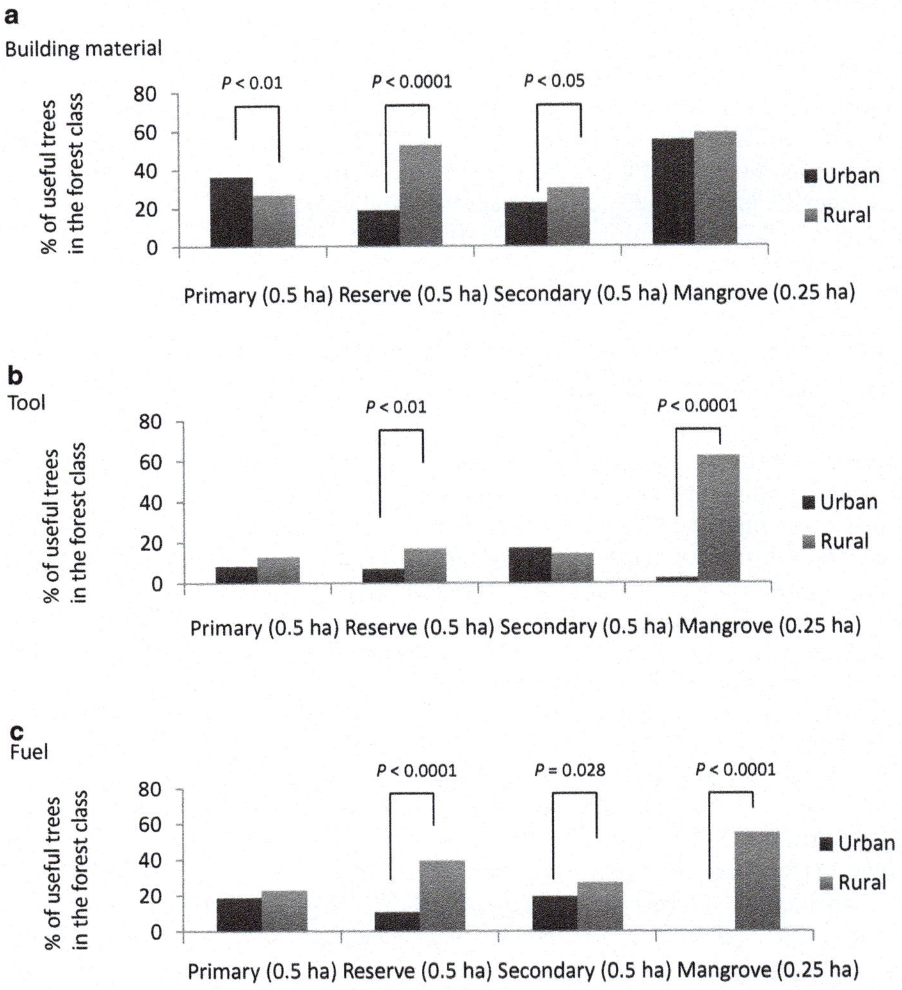

Fig. 5.8 Comparison of proportions of useful trees (used as building materials (**a**), tools (**b**), and fuel (**c**)) in each forest class based on the opinions of urban and rural villagers

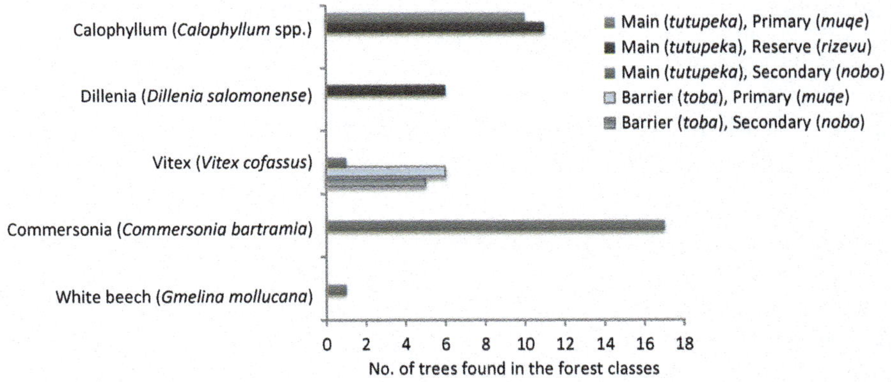

Fig. 5.9 Number of selected useful trees found in each forest class

important building material in the rural village, and white beech, a rare species. Vitex (*Vitex cofassus*), an important species used for tools and construction, was found on both islands, although it was more frequently observed on the barrier islands. These results suggest that each forest type has a different level of importance to the subsistence lifestyle of local peoples.

The biodiversity of the landscape was related to the people's use and management of forest resources. We found that the human-modified reserve and secondary forests represented distinct vegetative communities that are very different from both the primary forest, which has experienced few human impacts, and were also very different from each other (Table 3.8 and Figs. 5.1 and 5.2). For example, some pioneer species (e.g., *D. salomonense*) grew only in forest gaps in the reserve forest and not in mature forest; the villagers had created these gaps by the regular cutting of useful trees for their subsistence, which was permitted, although commercial use or clearance for gardens was not. Other tree species (e.g., *C. bartramia* and *Gmelina moluccana*) occurred only in secondary forest (Fig. 5.2), which regrew after shifting cultivation crop fields had been abandoned when productivity decreased.

Smith and Wishnie (2000) have also reported that moderate and repeated disturbance by other societies can enhance biodiversity at species, habitat, and landscape levels. Additionally, semidomestication is believed to have promoted the growth of specific species, such as Timonius (*Timonius timon*), which are uncommon in the wild but sometimes are encouraged to grow by the villagers when they do occur (Kobayashi et al. 2013; Matsui 1989). Thus, regular subsistence use of the forests by humans may have resulted in multiple unique vegetative communities, and higher overall biodiversity, than would otherwise exist (Furusawa et al. 2014). Alternatively, the diverse vegetative communities may have attracted people to this area; however, the small number of forest types near the more populous urban village contradicts this hypothesis. One may logically conclude that the existence of a variety of forest types, i.e., human-modified forests, both increases diversity and provides an essential base for the subsistence lifeways of rural people.

Was this human–biodiversity relationship intentionally constructed? Smith and Wishnie (2000) concluded that resource management and methods of sustainable use of traditional societies have commonly and indirectly resulted in conservation and the creation of biodiversity, although this had usually been unintentional. In practice, the villagers' activities such as the collection of forest resources from *rizevu* or semidomestication of plants were done to gain short-term benefits as a part of their subsistence lifeways rather than being designed for the long-term formation of biodiversity. Furthermore, as the Preface states, the villagers were aware of the relationships between different species in their local ecosystem. Therefore, the next chapter is devoted to exploring a cognitive model related to conservation and the use of biodiversity.

Chapter 6
Recognized Webs in Ecosystems

Abstract This chapter explores the psychological aspects of human–environmental/ecological interaction. The people of Roviana identified and acknowledged a variety of interactions that occur between humans, plants, and animals. They recognized the ecological fact that animals often affect humans through their relationships with plants, and these people recognized themselves as a part of the local ecological chains and food webs. However, this chapter also shows how the villagers recognized that people may have adverse effects on the ecosystem. For example, villagers reported their negative effects for plants they recognized as having direct negative effects on humans or, more simply, included plants they found problematic and typically killed. These analyses are used to discuss the existence of traditional conservation ethics that are designed to protect biodiversity and may affect it. In some ways, the people of Roviana are similar to people in Western countries, with an interest in short-term gains at the expense of the environment in some cases, while the conservation ethics of conservation organizations are often stronger than those of some individuals in both Roviana and in the West.

Keywords Folk ecology • Knowledge • Psychological domain • Human–plant–animal relationship • Indigenous ecological knowledge • Wild animal • Conservation ethics in recognition

6.1 Folk Ecology

Folk ecology and folk biology refer to a new interdisciplinary approach to understanding the everyday knowledge of the ecological or biological world of local people (Medin and Atran 1999). These methods are also being developed to address how knowledge is linked to action in diverse ecological and cultural contexts. These types of studies are fundamental to understanding how a society acts in a sustainable manner. In practice, anthropology (ethnobiology) and psychology (cognitive psychology) are combined in folk ecology and folk biology to learn how people understand and use interactions between plants, animals, and humans; this study focused on the cognitive model of individuals related to species relationships in ways that are relevant to environmental behavior (Atran et al. 1999, 2002).

T. Furusawa, *Living with Biodiversity in an Island Ecosystem*, Ethnobiology,
DOI 10.1007/978-981-287-904-2_6

This chapter explores the psychological aspects that underlie the human to natural world interactions in the community of Olive based on interviews. The goals are to address people's recognition of traditional environmental knowledge that they acquired through their relationships with the natural world or learned from their ancestors and to analyze that recognition and those relationships. By so doing, this analysis addresses two questions: (1) whether or not people recognize if their modification of the natural world may influence ecosystem webs and also recognize the ways in which other creatures directly or indirectly affect the environment and the people themselves and (2) whether or not people recognize the concept of conservation (as defined as the wise use of natural resources) as it relates to the plants and animals around them. Previous studies had doubted the existence of traditional conservation ethics in this region and worldwide among indigenous people and had criticized the view of "the ecologically noble savage" in two ways. First, those studies proposed that local people did not recognize the indirect benefits of their own actions on conserving or even improving local biodiversity. Those studies proposed that biodiversity in a human-modified environment is only a by-product of a "for-profit" subsistence lifeway. Second, previous studies have questioned whether local people had traditionally possessed "ethics" related to the conservation of biodiversity. That is why they tend to turn their resources, rights, and knowledge into commodities. Therefore, these two questions have very important implications for the study of the existence or nonexistence of conservation ethics.

6.2 Knowledge and Recognition of Nature

For the current study, lists of plants and animals were first developed to explore what the villagers thought of the world around them (Furusawa 2004). The plant list included 32 species with different ecological characteristics that humans related to in various ways (Table 6.1). The animal list included 28 animals, including 24 birds and four mammals, with different ecological characteristics and with various relationships with humans (Table 6.2). The next sections show the detailed criteria that were used for the selection and creation of the list of species.

Then during the next year, the main survey involved separate interviews with 14 individuals (eight males and six females)[1]; these individuals were not related within two degrees of kinship, did not belong to the same households, and included people of various ages over 30 years old. This group intentionally included people with a variety of knowledge and was carefully selected to avoid any bias based on close familial relationships, similar social rank, or by having only members in a single age group. The initial interview covered relationships between humans and plants. For every plant on the list, the following questions were asked: "Is this plant helpful/beneficial (*va leani/vinari tokae*) or harmful/disadvantageous (*ŋoŋovali*) to

[1] All individuals agreed to participate in this research after the author disclosed the methods and goals of the study.

Table 6.1 List of plants used during structured interviews and each species human–plant effects

Plant type	Common name	Scientific name	Roviana name	Use	Effect	
					Plant to human	Human to plant
Garden crops						
P1	Papaya	*Carica papaya* L	*Manioko*	Fruit/nut	+1	+0.93
P2	Banana	*Musa* spp.	*Hakua*	Fruit/nut	+1	+1
P3	Sweet potato	*Ipomoea batatas* L.	*Luzu vaka*	Tuber	+1	+1
P4	Cassava	*Manihot esculenta* Crantz	*Ore marihi*	Tuber	+1	+1
P5	Giant taro	*Alocasia macrorrhiza* Schott	*Karuvera*	Tuber	+1	+1
P6	Yams (introduced)	*Dioscorea* spp.	*Marihi*	Tuber	+1	+1
P7	Pana	*Dioscorea esculenta* L.	*Dikidiki*	Tuber	+1	+1
P8	Taro	*Colocasia esculenta* L.	*Talo*	Tuber	+1	+1
P9	Yam (indigenous)	*Dioscorea* sp.	*Lokete*	Tuber	+1	+1
Trees (including palms) grown near settlements and gardens						
P10	Java almond	*Canarium indicum* L.	*Okete*	Fruit/nut, medicine	+1	+1
P11	Ngali nut	*Canarium salomonense* B. L. Burtt	*Tovinia*	Fruit/nut	+1	+1
P12	Cutnut	*Barringtonia procera* (Miers) R. Knuth, *B. edulis* Seem.	*Kinu*	Fruit/nut	+1	+1
P13	Indian almond	*Terminalia catappa* L.	*Tatalise*	Fruit/nut, medicine	+1	+0.93
P14	Premna	*Premna corymbosa* Rottler & Willd.	*Zovi*	Medicine	+1	+0.14
P15	Coconut	*Cocos nucifera* L.	*Dohara*	Fruit/nut, medicine	+1	+1
P16	Betel nut	*Areca catechu* L.	*Heta*	Fruit/nut	+1	+1
Herbs/vines						
P17	Heartleaf hempvine	*Mikania cordata* (Burm. f.) B. L. Rob.	*Luluzu*	Medicine	–0.43	–1

(continued)

Table 6.1 (continued)

Plant type	Common name	Scientific name	Roviana name	Use	Effect	
					Plant to human	Human to plant
P18	Wedelia	*Wedelia rechingeriana* Muschl.	*Rabutu ŋira*	Medicine	+0.36	−0.86
Small-/medium-sized forest trees						
P19	Semecarpus forstenii	*Semecarpus forstenii* Blume	*Sakita*	Harmful	−1	−1
P20	Brown kurrajong	*Commersonia bartramia* (L.) Merr.	*Zamara*	Building	+1	−0.29
P21	Galearia celebica	*Galearia celebica* Koord.	*Zizito*	Medicine	+1	−0.07
P22	Dendrocnide nervosa	*Dendrocnide nervosa* (H. J. P. Winkl.) Chew	*Zilatoŋo*	Harmful	−1	−1
Large forests trees						
P23	Flueggea	*Flueggea flexuosa* Müll. Arg.	*Mavuana*	Building	+1	−0.14
P24	White beech	*Gmelina moluccana* Backer ex K. Heyne	*Qoliti*	Canoe	+1	+0.43
P25	Dillenia salomonensis	*Dillenia salomonensis* (C. T. White) Hoogl.	*Kapuhu*	Building	+0.86	−0.36
P26	Whitewood	*Endospermum formicarum* Becc.	*Kakadikana*	Harmful	−0.36	−0.64
P27	Vitex	*Vitex cofassus* Reinw. ex Bl.	*Vasara*	Building	+1	−0.50
P28	Calophyllum	*Calophyllum* spp.	*Buni*	Building	1	−0.14
Introduced timber plantation species						
P29	Rainbow eucalyptus	*Eucalyptus deglupta* Blume	*Iuki*	Cash	+1	+1
P30	Teak	*Tectona grandis* L.	*Tiki*	Cash	+1	+1
P31	Big-leaf mahogany	*Swietenia macrophylla* King.	*Suim*	Cash	+1	+1
P32	English beechwood	*Gmelina arborea* Roxb.	*Melaina*	Cash	+1	+1

Table 6.2 List of animals, 24 bird and four mammal species, discussed during structured interviews and the associated human–animal effects

Animal type	Common name	Scientific name	Roviana name	Life form	Effects	
					Animal to human	Human to animal
Predators						
A1	Solomon sea-eagle	*Haliaeetus sanfordi*	*Atata*	Bird	−0.14	−0.29
A2	Osprey	*Pandion haliaetus*	*Manuvu*	Bird	+0.14	0
A3	Brahminy kite	*Haliastur indus*	*Nae*	Bird	−0.14	−0.29
Birds common in forest edge habitat						
A4	Imperial pigeon	*Ducula* spp.	*Baruku*	Bird	+0.86	+0.79
A5	Melanesian cuckoo-shrike	*Coracina caledonica*	*Hidoko*	Bird	0	−0.14
A6	White-bellied cuckoo-shrike	*Coracina papuensis*	*Pisale*	Bird	0	+0.07
A7	Metalic pigeon, Yellow-legged pigeon, and/or Claret-breasted fruit dove	*Columba vitiensis, C. pallidiceps, Ptilinopus viridis*	*Kukuva*	Bird	+0.36	+0.29
Small birds common in open areas						
A8	Honeyeater	*Myzomela* spp.	*Hililiboe*	Bird	0	0
A9	Starling	*Aplonis* spp.	*Hiuheze*	Bird	0	+0.07
Forests animals						
A10	Nicobar pigeon	*Caloenas nicobarica*	*Bakupa*	Bird	+0.36	−0.21
A11	Dollarbird	*Eurystomus orientalis*	*Kilikora*	Bird	−0.14	−0.14
A12	White-winged fantail	*Rhipidura cockerelli*	*Pitikole*	Bird	0	−0.14
A13	Kingfisher	*Alcedo* spp.	*Tarabua*	Bird	0	0
A14	Yellow-faced myna	*Mino dumontii*	*Kinio*	Bird	−0.21	+0.07
A15	Cuscus	*Phalanger* spp.	*Manue*	Mammal	+0.36	−0.29
A16	Melanesian scrubfowl	*Megapodius eremita*	*Eo*	Bird	+0.36	−0.43
A17	Willie wagtail	*Rhipidura leucophrys*	*Vie*	Bird	+0.07	+0.14
A18	Mackinlay's cuckoo-dove	*Macropygia mackinlayi*	*Buti*	Bird	−0.21	−0.43
A19	Blyth's hornbill	*Aceros plicatus*	*Omehe*	Bird	−0.29	−0.07

(continued)

Table 6.2 (continued)

Animal type	Common name	Scientific name	Roviana name	Life form	Effects	
					Animal to human	Human to animal
Animals appearing in gardens						
A20	Ducorp's cockatoo	*Cacatua ducorpsii*	*Kakia*	Bird	−0.43	−0.21
A21	Singing parrot	*Geoffroyus heteroclitus*	*Kara*	Bird	−0.57	−0.36
A22	Rodent	Myomorpha	*Kurezu*	Mammal	−1	−0.64
A23	Yellow-eyed cuckoo-shrike	*Coracina lineata*	*Matakekeve*	Bird	−0.07	−0.07
A24	Pygmy-parrot	*Micropsitta* spp.	*Vilisuru*	Bird	+0.07	+0.21
A25	Cardinal lorry	*Chalcopsitta cardinalis*	*Siri*	Bird	−0.21	−0.07
Ground-dwelling animals						
A26	Purple swamphen	*Porphyrio porphyria*	*Balikuhu*	Bird	−0.86	−0.71
A27	Pig	*Sus scrofa*	*Boko*	Mammal	+0.43	+0.86
Flying mammals						
A28	Bat	Chiroptera	*Veke*	Mammal	−0.36	−0.57

you?" and "Do you help or harm this plant?" Answers were classified as yes, undecided, no, or "do not know this plant" and were coded as +1, 0, −1, and "not applicable," respectively. The author requested detailed reasoning for each answer. Negative human effects to a plant included human activities that negatively affected plant growth but excluded effects to mature trees (i.e., trees harvested for fuel and other subsistence-related needs). Thus, recognized actions and counteractions between humans and plants were measured.

The next interview covered plant–animal relationships. Using both the plant and animal lists, the first question was either, "Do you think this plant helps or harms this animal?" or "Do you think this animal helps or harms this plant?" The same coding system was used and detailed reasons for each answer were also requested. A final interview covered human–animal relationships. For each species on the animal list, each individual was requested to say if each animal helped/harmed that person and if he/she helped or harmed that animal, including detailed explanations.

For analyzing all the above relationships, the averaged value of everyone's answers, i.e., +1, 0, −1, while excluding "not applicable" responses, was calculated and is called the "effect" in this chapter. This effect thus ranged from +1 (all answered "helpful"), through 0 (neutral), to −1 (all "harmful").

6.3 Human–Plant Relationships

The author selected a variety of plants used in Olive village for further analysis; note that methods used for plants in this analysis and those used in Chap. 5 were similar. However, this study was conducted independently; the plant list used here and the one in Chap. 5 varied slightly. Thirty-two species were selected for the present study, including seven root crops, seven fruit/nut species, five species used to build houses, five species used to treat illness, and one species used to make a canoe as well as four introduced commercial timber species and three species known to be harmful to humans (Table 6.1). Obviously, this selection method biased this analysis to include only plants that humans used. Additionally, timber plantation species were also included. Species observed to have been planted during the research period included Eucalyptus (*Eucalyptus deglupta*), teak (*T. grandis*), big-leaf mahogany (*Swietenia macrophylla*), and English beechwood (*Gmelina arborea*). The author observed that the people had a negative view toward *Semecarpus forstenii* (*sakita*) and *Dendrocnide nervosa* (*zilatoŋo*), which caused severe skin inflammation, and whitewood (*Endospermum formicarum*: *kakadikana*), which, they claimed, killed neighboring trees. Although these three could be classified as the opposite of "useful," they were included because this study aimed to reveal people's recognition and activities related to various types of plants.

Table 6.1 also shows the recognized effects of several plants on humans and how humans affected these plants. Most effects of plants on humans were positive (30/32), although this was expected because 29 species were chosen from plants people found "useful." However, three plants were chosen from those known as harmful (i.e., *Semecarpus forstenii*, *Dendrocnide nervosa*, and whitewood) and had negative effects. Additionally, heartleaf hempvine (*Mikania cordata*), which was used to treat illness, was also recognized as having a negative (−0.43), harmful effect. This species, a creeping vine of open areas that grows rapidly enough to cover garden crops, is a main target for weeding. According to an elderly informant, this plant was an introduced species which initially invaded Munda around 1990 and rapidly spread throughout Roviana. This informant also insisted that while heartleaf hempvine was useful as medicine, it might have been introduced by a foreigner who was not from the local village. The villagers also considered it an invasive species.

In this study, humans were shown to have positive effects on the majority of useful plants, as was expected. Human effects were negative on only 11 plants, all undomesticated, which were all wild herbaceous plants and forest trees. Humans targeted these plants for weeding even in gardens or settlements; for example, all informants consistently rated heartleaf hempvine as −1 despite the fact that villagers commonly used this species for treating wounds. Villagers reported this species to be ubiquitous and said they could find this species anytime they needed them. Typically, villagers reported negative effects for plants they recognized as having direct negative effects on humans or, more simply, included plants they found problematic and typically killed. One elder said that it had become very rare to encounter the whitewood, which may have allelopathy, because they cut all whitewood trees as soon as they found them. White beech (*qoliti*: *G. moluccana*), an interesting exception, grows wild and received a relatively high positive effect (+0.43) from humans because, as explained in Chap. 5, it is very important as it is the only tree species used to make canoes and only rarely grows in natural forest gaps. That is, people found this tree so valuable that they were likely to help the tree grow even though white beech trees were weeded from gardens to protect crops from competition.

6.4 Human–Animal Relationships

Next, the analyses looked at human–animal relationships using the list of animals (Table 6.2). The fauna of the Solomon Islands is not as rich as on mainland Asia or even New Guinea Island because the Solomon Islands are isolated from those areas. Roviana mammals include pigs (Roviana name, *boko*), species of marsupials known as cuscus (*manue*: *Phalanger* sp.), various kinds of rodents (*kurezu*), and bats (*veke*) (Pikacha 2008). Pigs, considered valuable in traditional rituals, experienced the highest hunting pressure from humans as their primary hunting target. Villagers had also domesticated some pigs. At the time of fieldwork, the pigs were eaten infrequently only at ceremonial events (such as weddings) and also were sold to Asian logging company managers. However, Rappaport (1984) and other scholars have noted that pigs commonly have high ritual value in Melanesia, especially in Papua New Guinea societies. However, pigs did not seem to be as important in Roviana as reported elsewhere. While cuscuses and bats were also sometimes hunted and eaten, this was very rare. Humans rarely found common reptiles and amphibians useful, such as crocodiles (*basioto*), frogs (*bakarao*), and snakes (*noki*).

About 70 species of birds occur on New Georgia Island (Blaber 1991; Doughty et al. 1999). While this number is not very large, a high proportion of them are endemic. Eagles, hawks, hornbills, colorful parrots, lories, and kingfishers always attract the attention of tourists. Informants reported that in the past, people hunted birds with bows and arrows or with guns, but currently the use of guns is strictly banned so hunting has mostly been discontinued. In general, the relationships humans have with animals seemed superficial; nevertheless, people were observed to have a rich knowledge of the biological characteristics of various animals.

Humans rarely encounter terrestrial animals, so qualitative interviews were used to create the animal list rather than basing that list on some quantitative method. The author visited 12 village elders (six of each gender) and asked each to name all the animals that they could think of that positively or negatively affect plants or humans. They named five mammals, 32 birds, a few reptiles, and amphibians as well as a number of insects; four mammals and 24 birds were chosen for further analysis because they were considered to be commonly known by people (Table 6.2). The animals were classified into seven categories for convenience as follows: predators (3 species), birds common in forest edge habitat (4), small birds common in open areas (2), forest animals (10), animals appearing in gardens (6), ground-dwelling animals (2), and flying mammals (1).[2]

Nine animals had positive effects on humans (32 %), while 13 animals (46 %) had negative effects; the effects of others were neutral to humans. More animals than plants were recognized as harmful to humans, although this may have been caused by the method used to select species for the lists. Major reasons for positive effects of animals on humans included animals providing meat for people (pig, Melanesian scrubfowl, and cuscus) and animals becoming attached to people as pets (a variety of pigeons) (Fig. 6.1). The imperial pigeon showed the highest positive effect because, in addition to these reasons, this animal also dispersed seeds of useful tree species as is explained in detail in the next section. One informant even insisted that she trained this bird to get Java almonds or ngali nuts (*Canarium* spp.) in the forest and bring them to her.

Animals that that ate and destroyed garden crops or edible nut/fruit plants caused negative effects on humans. A variation of the absolute value of these types of negative effects (−1 to 0.07) from "animals appearing in gardens" depended on differences in body size and the magnitude of the destruction. For example, the very small body size of a pygmy-parrot and a yellow-eyed cuckoo-shrike caused informants to insist that their effects were minimal. Wild pigs also destroy root crops in gardens, although their negative effect was counterbalanced by their positive effect as a source of meat. Overall, many of the negative effects of animals were related

Fig. 6.1 An imperial pigeon interacting with humans

---

[2] In the actual analyses, cluster analysis was used to classify animals; impacts of plants and animals were used as variables for measuring the Euclidian distance, and the Ward method was used for clustering.

characteristically to the indirect negative effect that animals had on humans; these indirect effects occurred when animals destroyed useful plants rather than occurring as negative effects of animals attacking humans directly.

Humans effected animals in various ways. During the interviews, several informants observed that “harm to these animals is good because they may have bad effects on me” or “help provided to these animals is good because they may have made a contribution to me.” These ideas reflected the fact that they seldom had direct contact with the animals in real life. Generally, humans helped only imperial pigeons and pigs, which were recognized as providing reciprocal help to humans, while they eliminated other animals, did nothing about them, or revealed no interaction with some animals (neutral).

6.5 Plant–Animal Relationships

Finally, the plant–animal relationships recognized here were analyzed. Paired relationships between 32 plants and 28 animals show how plants effect animals (Fig. 6.2).

In almost all pairs, plants were recognized as helpful to animals or neutral. Plants provided foods (e.g., P1 (papaya) to A28 (bats), P10 and P11 (*Canarium* spp.) to A4 (imperial pigeons) and A19 (Blyth’s hornbills)), and shelter (e.g., P26 (*Dillenia salomonensis*) to A2 (osprey)) and provided nesting materials for animals; Table 6.1 provides the codes used in this paragraph for species plants (P-numbers) and animals (A-numbers). Villagers also observed more detailed interactions. For examples, root crops such as sweet potatoes (P3) and taro (e.g., P8) provided foods to pigs (A27) and purple swamphen (A26), because these animals live on the ground and dig up soil to find food. Java almond (P10) and ngali nut (P11) provided food for birds; interestingly, the villagers recognized that only those birds which had strong, heavy beaks could eat these foods and break the hard shells that covered these nuts (e.g., imperial pigeon (A4), Blyth’s hornbill (A19), and Ducorp’s cockatoo (A20)). The informants stated cutnut (P12) had positive effects (over +0.5) on only bats (A28) and rodents (A22) because its inflorescences hang vertically from branches; birds could not remove the nuts from the inflorescence, so cutnut had an almost neutral relationship with birds and other animals. Although *Semecarpus forstenii* (P19) and *Dendrocnide nervosa* (P22) were recognized as causing skin problems on animals as well as humans, their effects on animals were minimal, because, as most informants recognized, animals avoided these plants.

Figure 6.3 shows the effects of the same plants on animals, but these effects were classified into respective plant and animal groups (Tables 6.1 and 6.2). Villagers recognized all sizes of trees in these forests as providing shelter and nesting materials to animals. Unlike nut/fruit trees, these kinds of trees provided benefits to predatory birds (e.g., Solomon sea-eagle and osprey). Effects of plants which provide food for animals (e.g., garden crops) varied according to the animal categories; effects of trees which provide shelter to animals (e.g., large trees in forest) were

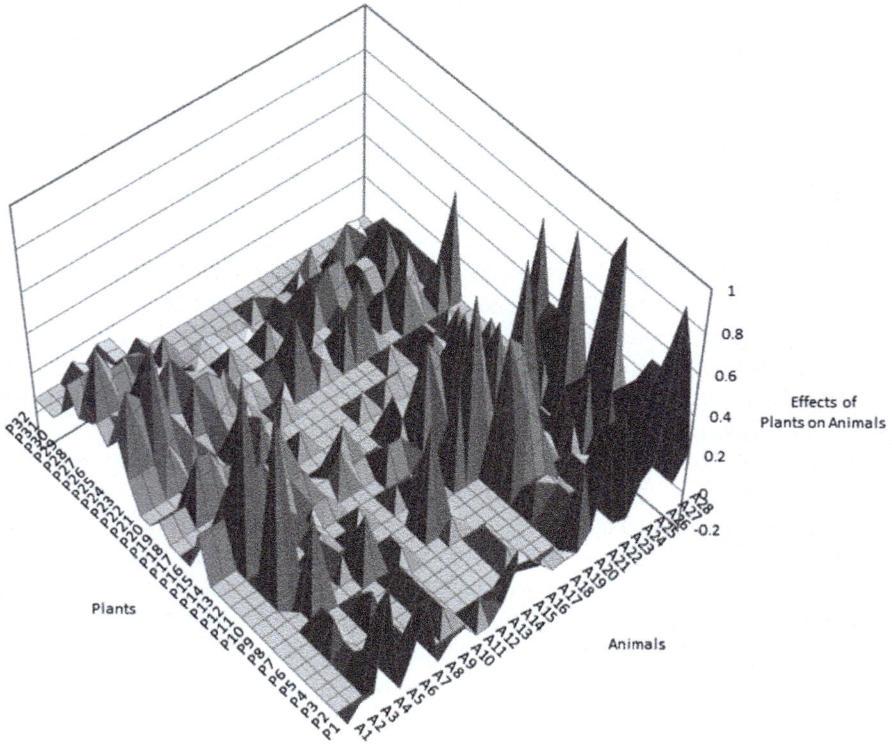

Fig. 6.2 Effects of plants on animals (See Tables 6.1 and 6.2 for corresponding plant and animal names along with their respective types)

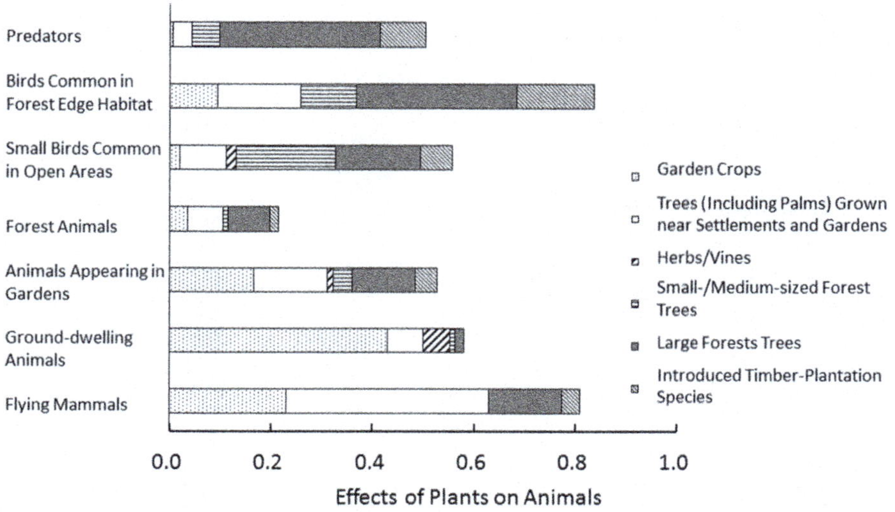

Fig. 6.3 Average effect of categories of plants on categories of animals

similar in the different animal categories (Fig. 6.3). The average positive effect of plants on animals was highest for "large forest trees," nearly 1.5 times larger than for "garden crops." Note that these "large forest trees" were also targets in recent commercial logging efforts; in reality, virtually all high effect trees were exploited by loggers, a point that will be discussed later in more detail.

Next, the effects of animals on plants were analyzed (Fig. 6.4) and the majority of non-neutral effects were negative primarily because herbivore animals eat and destroy nuts (P10 (Java almond) and P11 (ngali nut)) and fruits (P1 (papaya) and P2 (banana)). Additionally, a small number of positive effects were recognized (right side of Fig. 6.4). Very interestingly, imperial pigeon (A4) was recognized as having a positive effect because this bird liked to eat Java almond (P10) and ngali nut (P11), but their beaks did not destroy the seeds, allowing the seeds to be dispersed to other places (i.e., helping seed dispersal: effect>+0.05 to respective species). Other birds, such as Blyth's hornbill (A19), were recognized as harmful because their strong beaks destroy seeds through the seed's outer shell. Other positive animal effects on plants resulted from bird droppings enriching soils.

Analyses were also made of the categorized types of animals (Fig. 6.5). Effects were biased toward the negative (Fig. 6.4). "Ground-dwelling animals" and "animals appearing in gardens" had strong negative effects on "garden crops" and "trees (including palms) grown near settlements and gardens." In tandem with previous analyses of the effects of plants on animals, these animals received benefits from these plants that they also consumed; this relationship seemed to be "returning good

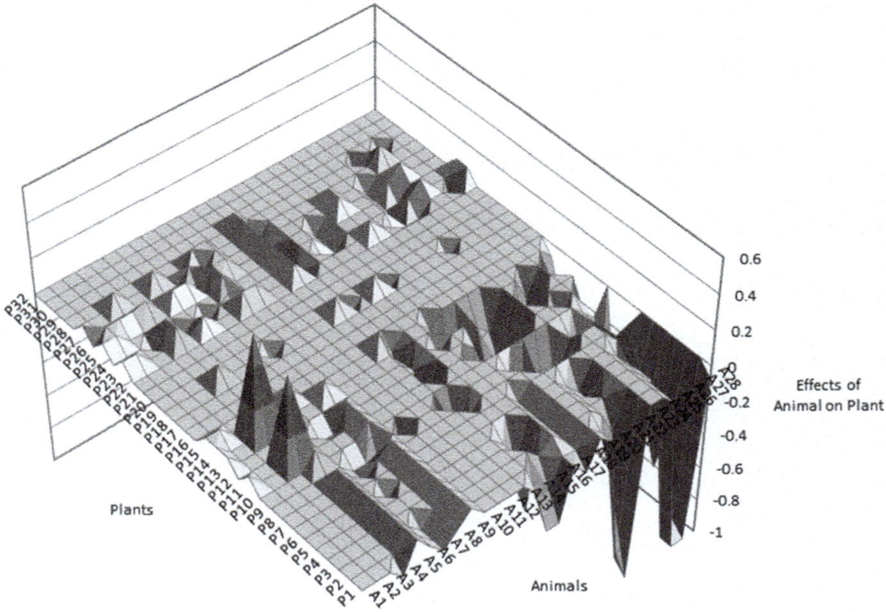

Fig. 6.4 Effects of animals on plants (See Tables 6.1 and 6.2 for the corresponding plant and animal names along with their respective types)

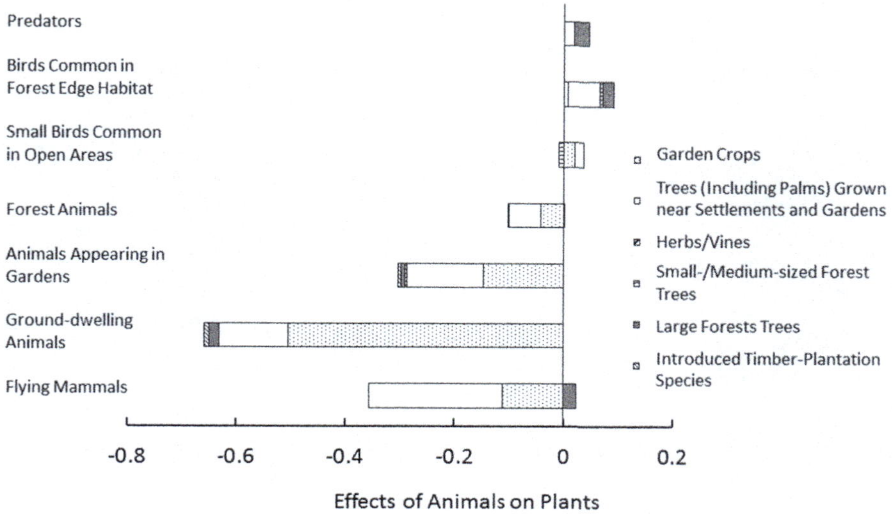

Fig. 6.5 Average effect of animal categories on plant categories

for evil" for these herbivorous animals. Even though "large forest trees" and "small-/medium-sized forest trees" had positive effects on "predators" and "birds common in forest edge habitat" animals, the positive effect of animals on these plants was minimal and this relationship seemed to be commensal. However, droppings of predators and such herbivores (the "birds common in forest edge habitat") were recognized by the villagers as enriching soils near trees in which these animals took shelter. Additionally, birds, such as imperial pigeon, helped disperse seeds of plants that they consume in a relationship that seemed to be reciprocally beneficial.

Finally, recognized ecological chains involving humans, plants, and animals are discussed based on analyses of "human–plant," "plant–animal," and "animal–human" relationships. The correlations between the "plant → animal → humans" chain dealing with the "effects of plants on animals" and "effects of animals on humans" were analyzed; the results (Spearman's rank correlation analysis) indicated no significant correlation. Then, the "animal → plant → human" chain was studied. Correlation analysis disclosed that a significant negative correlation exists between "effects of animals on plants" and "effects of plants on humans" ($\rho = -0.38$, $P < 0.05$). This suggests that animals adversely affected plants that humans found useful and those useful plants were more likely to be destroyed by animals than other plants, a relationship familiar to anyone who has been involved in gardening. Then, correlation between "effects of animals on plants" and "effects of animals on humans" was analyzed and showed a significant positive correlation ($\rho = 0.39$, $P < 0.05$); that is, animals that harmed plants were also harmful to humans (Fig. 6.6).

Obviously, the villagers recognized the ecological fact that animals often affect humans through their relationships with plants and these people recognized themselves as a part of the local ecological chains and food webs.

Fig. 6.6 Examples of recognized human–plant–animal relationships. *Arrows* indicate positive effects

6.6 Development and Changing Recognition of Ecological Webs

The findings in this chapter can be analyzed in relation to recent developments. First, people in the Saikile customary land (including Olive) had signed an agreement with the logging company; in this agreement, in addition to the Forestry Act, loggers were not allowed to harvest locally important species. The plant list in this chapter identified six tree species with high commercial value, including two traditionally important species that were identified as being protected: ngali nut (*C. salomonense*) and white beech (*G. moluccana*). Logging companies were allowed to harvest the remaining four species (*D. salomonensis*, whitewood, vitex (*Vitex cofassus*), and calophyllum).

Villagers recognized that the effects of protected species on humans were all positive (+1); furthermore, those of species harvested by loggers were rated at +0.63, suggesting all these species were recognized as being very useful. In contrast, effects of humans on the two protected species listed above averaged +0.71, while those on species harvested by loggers averaged −0.41. That is, while people recognized that all six species provided them with benefits, they helped the growth of two protected species but harmed four unprotected species by harvesting their logs. Both types of tree species, those protected from harvest and species harvested for logging, had positive effects on animals ranging from +0.12 to +0.19 for the logged species and +0.16 for both protected species. Thus, the species harvested for logging were also recognized as useful to both humans and animals, yet were subject to weeding. As mentioned in Chap. 5, Roviana people had not been able to cut and use large trees, such as vitex and calophyllum, until the arrival of the iron axe and chainsaw. In reality, the forest had been an awkward resource for them to use because they needed to spend a large amount of labor to clear the forest to make way for productive gardens; clearing a forest still involves a considerable amount of hard work, even for a secondary forest. The logging operations helped the people by removing a number of large forest trees; the loggers also built a road into the inland forest. The people found it reasonable to welcome logging because it provided them more opportunities to cultivate new gardens.

In accepting commercial logging, the people might find a balance between the benefits (money paid for logging rights and new infrastructure and by making horticulture easier) and losses (e.g., trees). The analyses of logged species and other tree species in this chapter suggest that the people tended to put less importance on uncultivated species; these species were targets of weeding. Another implication was that wild animals were generally recognized as harmful; so, the adverse effects of logging on wild animals were also acceptable to the villagers. In this sense, commercial logging provided people with a significant economic benefit while exploiting both non-important and even harmful trees species, as long as the company avoided harvesting the protected tree species. However, ecologically speaking, this was not always true. When the shallow soils of a tropical rainforest ecosystem are exposed to strong sunshine, these fragile soils are easily erodible; furthermore, the original ecosystems had high levels of biodiversity. The people may lose invaluable ecosystem services when the forests are harvested.

The local people have invested a considerable amount of time and effort in developing timber plantations, also called reforestation by the local people, as a potential source of future cash income. The people were likely to clear a larger area of forest with the goal of generating a larger opportunity for future cash income. The effects from introduced timber species were all +,1 and human effects were the same, even though people had not harvested and earned money from these forests at the time of this study (Table 6.1). This pattern was symbolic and stood in contrast to the negative human effects on uncultivated plants. Plantation forests were recognized as requiring the care and attention of people, while natural forests can grow naturally and independently of human effects.

6.7 Conservation in Recognition

Analyses were also made from the viewpoint of the recognitive background of conservation. Even though the villagers welcomed the logging activities because they provided an opportunity for cash income, this did not necessarily mean the villagers lacked conservation ethics, because the people strongly protected and helped the growth of traditionally protected trees. Recently, the government officers recommended that the local people protect and help the growth of vitex and calophyllum because these trees were now both vulnerable to logging and would become a future source of logging income. The villagers did not yet follow this recommendation, showing an apparent lack of conservation ethics. However, time may be necessary to judge the real necessity of protecting these species. Importantly, when one considers the attitude of the villagers to the introduced timber species, a single community may have both economy-oriented resource exploitation and a moral desire to support conservation as sometimes occurs in typical Western societies.

Another important finding was that the local people were able to recognize that their direct modification of the environment (e.g., effects on plants) might influence other creatures because human effects on plants in turn influence the effects of plants on animals. That is, ethics are apparently lacking because people are willing to harm even useful floral/faunal resources and therefore exhibit a lack of ethics from a Western conservation viewpoint. In depth, however, the people's decisions and actions are rooted in more complicated values in ecosystems and they protect some species and vegetation. A type of "unintentional" conservation/creation of biodiversity may be a product of "intended" recognition. These findings suggested that conservation ethics is also diversified within a community or even within an individual, because each individual may be inclined to exploit resources and clear a forest for short-term gain without listening to recommendations of outsiders while they continue to conserve traditionally important species and recognize the complicated nature of the local ecosystem. In some ways, the people of Roviana are similar to people in Western countries, with an interest in short-term gains at the expense of the environment in some cases, while the conservation ethics of conservation organizations are often stronger than those of some individuals in both Roviana and in the West.

Chapter 7
Changing Ethnobotanical Knowledge

Abstract This chapter examines the acculturation of individuals to ethnobotanical knowledge in association with the effects of modernization. I examined the effect of modernity, which was measured as a first principal component; the variation in this principal component was tied to the cash economy, demographics, and contacts with Western materials. Complex intervillage differences in knowledge were correlated with modernity, in complex ways, as follows. The differences in knowledge were correlated with modernity when the villages were referenced to the least-modernized village. Contrary to our expectations, this did not occur when villages were referenced to the village with the highest modernity score. These results implicated that modern Western knowledge is easily integrated into the ethnobotanical knowledge system and is related to the loss of diverse botanical knowledge. An increase of Western knowledge may bring a more accurate level of biological knowledge to the villagers, but this new knowledge may not be adaptive to those living in the local environment. Even worse, Western knowledge may be biased in that villagers only gain knowledge related to how they can gain short-term for-profit-related benefits.

Keywords Modernization and loss of ethnobotanical knowledge • Modernity score • Acculturation • Cultural consensus • Cultural competence • Diversity of knowledge

7.1 Changing Ethnobotanical Knowledge

Previous chapters analyzed the behaviors of humans as they related to plants as well as their knowledge of those plants. However, several reports indicate that modernization has resulted in a loss or reduction of the ethnobotanical knowledge among indigenous peoples (Case et al. 2005; Lewis 2003; Reyes-Garcia et al. 2007; Romanucci-Ross et al. 1997). For our purposes, the term "indigenous people" refers to the indigenous people of Roviana who are discussed in this book in the sense of being a local and preindustrial society that is gradually being more modernized over time through interaction with the industrial world. If traditional ethnobotanical

T. Furusawa, *Living with Biodiversity in an Island Ecosystem*, Ethnobiology,
DOI 10.1007/978-981-287-904-2_7

knowledge is lost, the positive effects of the knowledge and related activities that sustain or enhance biodiversity may be reduced. Previous studies suggest that factors influencing this loss include distance from towns and the availability of natural resources (Case et al. 2005; Prance et al. 1987; Reyes-Garcia et al. 2007), the integration of Western materials and market-oriented knowledge into tribal communities (Bennett and Prance 2000; Benz et al. 2000; Ragone et al. 2001; Zent 2001), as well as age, gender, and ethnicity (Atran et al. 2002; Boster 1986; Caniago and Stephen 1998; Case et al. 2005). This process is similar to the process of revealing local knowledge to new members in a society.

This chapter addresses how knowledge has been formed and is transforming Roviana under current ongoing socioeconomic changes. The people of Roviana, like other populations in the Solomon Islands and Melanesia in general, share a similar cultural background and ecological settings. Furthermore, even within a population, the degree of modernization—i.e., progress related to a cash economy, population expansion, availability of modern goods and infrastructure, and exposure to Western culture—varies from one village to another. These situations indicate that individuals and villages within each language group have experienced various levels of modernization. For this purpose, this chapter adopted quantitative methodologies to measure how modernization has changed human ethnobotanical knowledge, using a cultural consensus model (Atran et al. 1999; Reyes-Garcia et al. 2003; Romney 1999; Romney et al. 1986).

7.2 Cultural Consensus and Modernity

7.2.1 Conceptual Framework

Cultural consensus analysis and the calculation of cultural competence have been developed as methods that have proved to be suitable for evaluating "knowledge." These methods calculate the similarities and differences in how knowledge varies between or within populations and to what extent each informant comprehends a particular bit of knowledge or understands the knowledge of a community as a whole (Atran et al. 2002; Romney 1999; Romney et al. 1986). Statistical analyses can be performed to detect relationships to differences in other variables such as modernity by quantifying the similarities of knowledge between individuals.

The concept of cultural consensus analysis involves revealing whether or not a single factor applies to all informants and to what extent that factor affects each individual. The culture is analyzed as a knowledge base that is both shared and learned by the individual members of the culture. This allowed development of a cultural consensus theory that can be used as a method to describe and measure the extent to which the informants share specific cultural beliefs. The model assumes that the correspondence between the answers of any two respondents is a function of the extent to which each answer is correlated with some truth. The single factor, which is a mathematical product that is virtually referred to as culture, is assumed to exist when the ratio of eigenvectors of the first principal component to the second is markedly large (more than three times), the proportion of the first eigenvector explains most of the variation, and the factor loadings on all questions are positive, as a result of the analysis of an adjusted informant-by-informant agreement matrix (Borgatti 1996). The informant's cultural competence is the factor-loading score of the first principal component. The word knowledge refers virtually to the informants' responses to the ethnobotanical questions, but cultural competence, which is the extent to which each informant's knowledge coincides with the culture (single factor), is a good measure, because it is a score from which random effects and between-question bias have been removed.

The methods of analysis used here allowed the examination of acculturation in ethnobotanical knowledge and its correlation with modernization. First, the possible existence of a shared ethnobotanical culture in Roviana is examined. This provides the basis of this study, because the Roviana people are assumed to have a single cultural background and origin as well as to have experienced similar acculturation activities (Reyes-Garcia et al. 2003). Second, modernization caused changes in ethnobotanical knowledge in the correlative relationship. In the working hypothesis, this can be analyzed as the similarities or dissimilarities of knowledge between any two villages correlated with the differences in modernity between them. The average value of the adjusted agreement of answers of all pairs of informants between two villages was used as the measure of knowledge similarities. Note that this is mathematically similar to the average cultural competence of all informants in the paired villages. Additionally, differences in estimated modernity scores between two paired villages are used as measures for differences in modernity, to allow the analysis of the correlation between the similarity/dissimilarity of knowledge and that of modernity. Note that the change in knowledge is reflected as the similarity of village informants' knowledge to that of villagers in the least-modernized village. In the cultural consensus analysis, similarities of knowledge are measurable, but villagers' knowledge itself is not measurable unless referenced to a base village.

This chapter reports two phases of data collection: panel research[1] and cross-sectional research.[2] Panel research was used to study Dunde, a modernized village located near the township of Munda, and Olive, a less-modernized village; see Chaps. 2, 4, and 5 for a detailed explanation on these two villages. A total of 17 (114 members) and 15 (112 members) randomly selected households from Dunde and Olive villages, respectively, accepted the offer to participate in this study. In the panel research, the author observed plant-use behaviors, collected data on these behaviors, and these data were used to create a list of plants the Roviana people found useful, in addition to demographic, socioeconomic, and ecological data at the individual and household levels. As a result, 149 local vernacular plant names were listed, of which 98 were used in both villages. Of the plants used in both villages, 20 plants were randomly selected and structured interviews of the participants provided detailed information related to plant use (Table 7.1).

[1] From July to November 2003.

[2] From December 2003 to February 2004.

Table 7.1 List of plants used in the interview for cultural consensus analysis and measuring similarity and dissimilarity of knowledge within and between villages

Common name	Scientific name	Roviana name	Uses observed [a]
Bamboo	*Bambusa* spp.	*Beti*	B, T
Pemphis	*Pemphis acidula* J. R. & G. Forst.	*Bobogele*	B, T
Calophyllum	*Calophyllum inophyllum* L./*C. kajewskii* A. C. Sm./*C. vitiense* Turill	*Buni*	B, T
Watermelon	*Citrullus lanatus* (Thunb.) Mansf.	*Deri*	Fd
Star fruit	*Averrhoa carambola* L.	*Opiti*	Fd
Giant swamp taro	*Cyrtosperma chamissonis* (Schott) Merr.	*Gorehe*	Fd
Mint	*Mentha arvensis* L.	*Habe*	M, T
Dillenia	*Dillenia ingens* Burtt	*Hebere*	B, Fl
Syzygium	*Syzygium* sp.	*Hipahipala*	B, Fl
Malay apple	*Syzygium malaccense* Merr. et Perry	*Hipala*	Fd
Jackfruit	*Artocarpus heterophyllus* Lam.	*Beta*	Fd
Palaquium	*Palaquium erythrospermum* Lam.	*Paloto*	B, Fl
Pineapple	*Ananas comosus* (L.) Merr.	*Paenapolo*	Fd
Pumpkin	*Cucurbita moschata* Duchesne	*Pamuken*	Fd
Fijian longan	*Pometia pinnata* Forst. f.	*Qema*	B, Fl, T
Teak	*Tectona grandis* L.	*Tiki*	T
Ngali nut	*Canarium salomonense* Burtt	*Tovinia*	Fd, M
Vitex	*Vitex cofassus* Reinw. ex Bl.	*Vasara*	B, Fl, T
Galearia	*Galearia celebica* Koord.	*Zizito*	M
Premna	*Premna corymbosa* (Burm. F.) R. & W.	*Zovi*	B, M, T

[a]Observation of dietary habits, garden crops, housing styles, treatment for illness, collection of fuels, and tools owned in Dunde and Olive villages; *Fd* food, *B* building material, *M* medicine, *Fl* fuel, and *T* tool

The villages of Tombo, Ha'apai, Nusa Banga, Mandou, and Rarumana were selected for cross-sectional research, in addition to Dunde and Olive (Fig. 7.1). These seven villages exhibited different stages of modernization within Roviana.

7.2.2 *Cultural Consensus*

For cultural consensus analyses, interview-based surveys were conducted. Only six households from Tombo, a small village with eight households, participated because the members of two households were absent during the study period. Ten of the 20 households were randomly selected from Ha'apai, and ten were also randomly selected from both Nusa Banga and Rarumana villages. In Mandou, 12 households were randomly selected. For each household, if either the male or female head (e.g., male head or his spouse, widowed female, or her first son) was originally from outside the village, the other person was automatically selected as the informant, while,

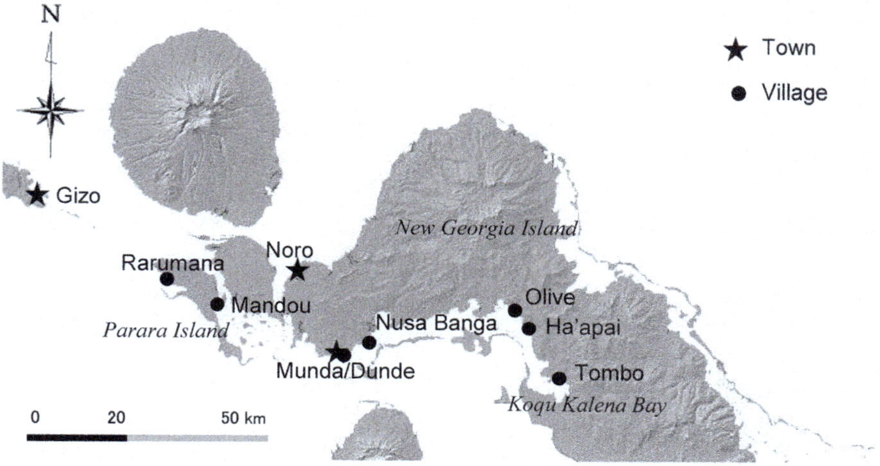

Fig. 7.1 Locations of the seven villages visited for a cross-sectional study of changing ethnobotanical knowledge

if both were from the same village, only one was selected at random.[3] As mentioned above, 17 and 15 households were randomly selected in Dunde and Olive, respectively, in the panel research. Thus, several interviewed informants represented each of the seven villages (80 total informants). Every participant in the cross-sectional research was asked to say whether each plant was used for (1) food, (2) building, (3) medicine, (4) tools, (5) magic/rituals, or (6) cash income. The informant could attribute more than one use to each plant. Cultural consensus analyses were performed according to standard procedures (Romney et al. 1986) for each village, each pair of villages, and for the total of all villages.[4] The similarities of knowledge between two villages were defined as the average of the adjusted value of coincident responses. The cultural competences of informants in the panel-research villages were calculated from all the three kinds of analyses listed above: (1) competence in each village's culture, (2) in culture shared by both villages, and (3) in culture shared by all Roviana informants.

A cultural consensus was observed in each of the seven participating villages (ratio of first to second eigenvectors was 13.2), in each pair of villages (average ratio was 11.8), and for all informants (ratio was 11.4; Table 7.2). The first eigenvector explained more than 80 % of the variation and was positive for all informants indicating a high degree of knowledge consensus. The degree of consensus varied among villages (from 6.0 for Nusa Banga to 21.8 for Olive) and among pairs of villages (from 6.9 for Dunde–Tombo to 16.9 for Olive–Tombo). Table 7.2 also shows the averages and ranges of individual cultural competences varying from 0.75 to 0.83.

[3] Note that at least one of the heads was born in the village, because, customarily, couples from outside the village do not hold rights to use the land and those people are not allowed to settle in the area.

[4] All cultural consensus analyses were performed using ANTHROPAC 4.0 (Analytic Technologies).

Table 7.2 Intravillage and intervillage cultural consensus [mean and range in block parentheses (minimum, maximum) are shown]

	N	Ratio first/second component	Consensus (%) explained by first component)	Average informants' competence[a]	SD of informants' competence
Intravillage	7	13.2 [6.0, 21.8]	89.5 [83.3, 95.6]	0.81 [0.75, 0.83]	0.04 [0.04, 0.05]
Intervillage	21	11.8 [6.9, 16.9]	88.2 [84.6, 91.0]	0.81 [0.78, 0.83]	0.05 [0.03, 0.07]
Overall	1	11.4	88.8	0.80	0.05

[a]Informant's cultural competence is the factor-loading score of the first principal component

7.2.3 Modernity Score

The other quantitative variable, a modernity score, is defined as the first principal component of the different kinds of variables that measure modernity in this study (Perz 2003): the intercorrelations among different variables are adjusted by calculating principal components. Several scholars have recognized that the ongoing modernization of Melanesia is strongly related to the cash economy, population growth, and the availability of modern materials and infrastructure, such as schools, aid posts, and radiophones; population growth resulted from improved hygiene and nutrition and the size of the potential economic market (Dennett and Connell 1988; Hodge et al. 1995; King and Collins 1989). In this study, the modernity score is a first principal component of the variables relevant to demographic and economic expansion and reflects the extent to which the community is exposed to Western materials and infrastructure. The validity of a factor as a proxy of modernity in this study is assumed when all the variables are positively correlated with the first component and when the factor explains most of the variation. Table 7.3 shows the variables used to reflect modernization events.[5]

The results show that all modernity variables for Dunde village, near the township of Munda, exceeded all such variables of other villages. Many of the more than 100 households of Rarumana village (22.3 km from the provincial capital of Gizo) possessed rainwater tanks and chainsaws. Ha'apai (32.7 km to Munda) and Mandou (11.8 km to Noro town) had 20 and 105 households, respectively, and modernity of these two villages was at intermediate level. Although Nusa Banga is near Munda town (7.7 km), the villagers there had fewer goods and less infrastructure (e.g., aid post, primary school, radios, and phones) than other villages. The villages of Olive and Tombo, which are far from Munda (31.7 km and 38.7 km, respectively), also had fewer possessions and these villages had less infrastructure. Principal compo-

[5]In calculating the modernity score, categorical variables were quantified and then used as variables for principal component analysis. The first eigenvector was defined as the modernity score when the first eigenvector was large and positive to all modernization variables but negative to the distance to the nearest town. The PROC PRINQUAL procedure of SAS 9.1 (SAS Institute, Inc.) was used for quantification of categorical variables.

Table 7.3 Modernity features of the seven villages analyzed in this study

	Dunde	Rarumana	Ha'apai	Mandou	Nusa Banga	Olive	Tombo
Number of households	202	103	20	105	64	65	8
Year of village foundation	1891	1915	1952	1916	1957	1972	1972
Distance to the nearest town (km)[a]	1.39	22.31	32.74	11.78	7.73	31.73	38.71
Number of modern-style houses per HH[b]	0.86	0.74	0.85	0.50	0.30	0.28	0.63
Number of rainwater tanks per HH	0.93	0.96	0.30	0.57	0.36	0.17	0.38
Number of outboard motors per HH	0.33	0.28	0.45	0.17	0.17	0.11	0.25
Number of chainsaws per HH	0.24	0.16	0.10	0.04	0.05	0.05	0.13
Number of generators per HH	0.08	0.07	0.05	0.04	0.03	0.03	0
Number of stores per HH	0.05	0.07	0.05	0.04	0.02	0.03	0
Water line (yes=1; no=0)	1	0	1	0	1	0	0
Aid post (yes=1; no=0)	1	1	0	0	0	0	0
Primary school (yes=1; no=0)	1	1	1	1	0	1	0
Radiophone (yes=1; no=0)	1	1	1	1	0	0	0
Development project (yes=1; no=0)[c]	1	1	1	0	0	1	0
Modernity score[d]	4.36	2.66	0.35	0.11	−2.09	−2.45	−2.94

[a]Distance to the provincial capital (Gizo), port town (Noro), or airfield (Munda), measured using a geo-referenced Landsat ETM+ image

[b]HH, household

[c]Rural development projects by nongovernmental organizations or foreign governments (e.g., EU and Taiwan)

[d]Score in first principal component; non-metric variables were transformed using the PROC PRINQUAL procedure of SAS software

nent analysis showed that the first eigenvector explained 58.1 % of the variation and was positive for all variables (negative to distance); this factor is hereafter called the "modernity score," shown in the last row of Table 7.3. The difference of modernity between the most-modernized village, Dunde, and the second most-modernized village, Rarumana, was calculated to be 1.70 and that between Dunde and Tombo was 7.30.

7.3 Correlation Between Knowledge and Modernity

Figure 7.2 shows the relationships between the similarities of knowledge and the differences in modernity. In Fig. 7.2(a), Dunde, the most-modernized village, was set as the base village. Other reference villages were projected on their relative similarity or difference from Dunde. In this case, the correlation was weak and not significant ($\rho = -0.36$). Second, in Fig. 7.2(b) where all villages were referenced to the base value of Tombo, the least-modernized village, a clear negative correlation was observed between knowledge and modernity ($\rho = -0.095$, $P = 0.0032$). These results suggested that when the least-modernized village was the base, knowledge changes occurred in correlation with the progress of modernity. The knowledge did not correlate positively with modernity, as shown in the case when the most-modernized village was used as the base.

Table 7.4 shows all the correlation coefficients between the similarities of knowledge and differences of modernity. As the left column shows, a significant correlation was observed only when Tombo was used as the base village. However, as the right column shows, if the coefficients were calculated only for the correlation of villages with modernity scores higher than the base village, significance was

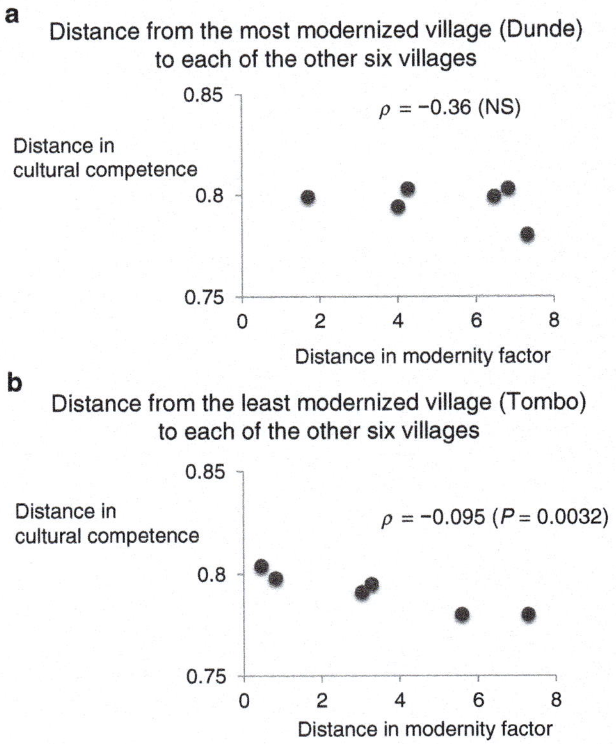

Fig. 7.2 Correlation between distance in modernity factor and distance in cultural competence

Table 7.4 Spearman's correlation coefficients between intervillage competence values and intervillage distance of modernity

Baseline village	Original ρ coefficient (N)	Coefficient ρ of correlation only with higher-modernity villages (N)
Dunde	−0.36 (6)	N.A. (0)
Rarumana	−0.32 (6)	N.A. (1)
Ha'apai	−0.48 (6)	N.A. (2)
Mandou	−0.38 (6)	−0.54 (3)
Nusa Banga	−0.52 (6)	−0.99 (4) $P=0.0055$
Olive	−0.40 (6)	−0.93 (5) $P=0.0203$
Tombo	−0.95 (6) $P=0.0032$	−0.95 (6) $P=0.0032$

Comparison between the original coefficients and coefficients when villages with a modernity score lower than the base village was excluded

observed in all cases of $N>3$. For example, with Nusa Banga as the base, the correlations between knowledge similarities and modernity differences were significant for Mandou, Ha'apai, Rarumana, and Dunde ($\rho=-0.99$, $P=0.0055$) but were not significant when Olive and Tombo were added. Although not shown in the table, a significant correlation was not observed when the analyses were performed only for villages with lower modernity scores.

Table 7.5 shows the number of plants known to be useful for each purpose. Villages were sorted by their modernity scores. For all uses other than cash income, the modernity score of the village did not correlate with the number of known plants. For example, Ha'apai villagers, who belonged to the village with the third-highest modernity score, mentioned a greater number of medicinal plants (7.5 out of 20 plants on average) than other villages. Meanwhile, the informants of Rarumana village, which had the second-highest modernity score, knew a greater number of plants used for tools (10.5 on average) than the less-modernized Nusa Banga (7.0). However, the effects of modernization were observed in the use category of cash income because the number of plants known as sources of cash income was correlated with the village modernity score ($\rho=0.85$, $P<0.05$).

7.4 Effects of Individual Characteristics

In the panel research (Olive and Dunde), male and female household heads, one of which was a participant for the interviews related to plants, were interviewed about individual and household information.[6] Individual knowledge is defined as cultural

[6] First, the dates of births and marriage (beginning years of the households) were obtained; official documents of birth and marriage were referred to during the interviews. Additionally, consumer unit (CU) was defined as the sum of each member's energy requirements based on body weight (Chap. 4); the producer unit (PU) was defined as the number of adults (18 years of age or older) in the household. Every evening, participating households were visited, and all members were asked

Table 7.5 Number of plants known to be useful for each of the seven use categories (mean ± S.E.), categorized by village. The villages were sorted in the order of decreasing modernity score [a, b, c]

	N	Food	Building	Medicine	Fuel	Traditional	Tool	Cash
Dunde	17	9.4 ± 1.1	8.5 ± 3.4	5.5 ± 3.4	11.4 ± 2.5†	4.9 ± 3.0	9.0 ± 2.5	16.4 ± 2.1
Rarumana	10	9.6 ± 1.3	9.1 ± 3.3	6.1 ± 2.4	12.1 ± 1.9	6.5 ± 2.1	10.5 ± 2.8#	16.2 ± 2.3
Ha'apai	10	9.7 ± 1.3	8.2 ± 2.5	7.5 ± 2.5#	12.7 ± 1.6	5.7 ± 2.2	10.3 ± 2.6#	16.2 ± 1.9
Mandou	12	9.8 ± 0.8	8.7 ± 2.2	4.0 ± 1.5†	12.8 ± 1.9	4.0 ± 2.1	9.2 ± 2.3	15.7 ± 2.2
Nusa Banga	10	8.5 ± 0.5	7.6 ± 2.7	6.1 ± 1.4	12.2 ± 2.0	3.6 ± 1.8	7.0 ± 2.4†	14.8 ± 1.3
Olive	15	9.1 ± 1.4	10.2 ± 2.3	6.3 ± 2.1	13.9 ± 1.1#	5.1 ± 2.6	10.1 ± 1.7#	15.5 ± 2.1
Tombo	6	8.7 ± 0.5	7.3 ± 3.0	7.0 ± 3.6	11.0 ± 3.6	6.0 ± 2.6	9.2 ± 3.1	15.3 ± 2.1

[a]# larger than † at $p < 0.05$ level, as evaluated by Tukey's multiple comparison for intervillage differences
[b]See Table 7.2 for the modernity score of each village
[c]Modernity score and the average number of useful plants were positively correlated with the use category of cash income ($\rho = 0.85$, $P = 0.015$) but not for other categories (ρ ranges from −0.29 to 0.64, $P > 0.12$)

competence (factor loadings of each informant; Borgatti 1996; Romney 1999). For this purpose, a model including the following factors was employed[7]: (A) individual-level basic variables (e.g., gender, age), (B) household-level life-stage variables (e.g., years after the marriage of the head of household and the number of consumers and producers), (C) variables of modernization at the household level (e.g., cash income), and (D) ecological variables (e.g., vegetation).

Multiple regression analysis was performed, after stepwise selection, to determine which individual and household characteristics affect individual cultural competence (see Table 7.6 for a full list of variables).[8] For the multiple regression analyses, three kinds of cultural competence values were used. Model 1 indicated the cultural competence of an individual compared well to the knowledge shared by all Roviana informants. For model 2, the cultural competence of an individual was compared with the knowledge shared by informants in the panel-survey villages (Dunde and Olive). Model 3 used the cultural competence of an individual compared with the knowledge shared by informants of the village where the individual lived.

Finally, individual-level knowledge was examined. Based on the results of a multiple regression analysis (Table 7.6), the cultural competences of the panel-survey villages (Dunde and Olive) were explained only by the cash income of the household, when competence was calculated as the knowledge shared by all Roviana informants (model 1). In analyses using competence for knowledge shared by informants in Dunde and Olive but not in the other villages (model 2), the detected factors were village, duration of the marriages of the heads, housing style, and informant's age. Higher cultural competence was observed for informants in Olive village, as well as those in younger households, for older informants, and for those who lived in Western-style permanent houses. In models 3-1 and 3-2, the cultural competences of the knowledge of individual Olive and Dunde informants, respectively, were used as dependent variables. The results show that age and living in a Western-style permanent house were positive factors indicating higher competences in Olive, but no significant factor relating modernity to knowledge was observed for Dunde village.

The effects of cash income on the competences of informants in the pool of all Roviana knowledge were reasonable, as an effect of the modernization described above. Exposure to a cash economy might affect the economic domain of botanical knowledge. The effects of the village when only the competence in the pooled knowledge of Dunde and Olive was used as the dependent variable were also reasonable, because villages represent different ecological settings and levels of mod-

---

to report all cash income and its source for 24 h during a 28-day period. Net income was calculated based on the amounts earned and the amounts expended for necessities. Everyday for 14 days, the participating households were visited every 1–2 h from 6 AM to 10 PM, all crop yields brought to the households were measured, and the total energy obtained per day was calculated in MJ (Chap. 4 for detailed methods). During the same period, the gardens cultivated by each household were visited and measured using a tape measure and clinometer.

[7] This model is based on a model proposed by Reyes-Garcia et al. (2007).

[8] Both significant level of entry (SLE) and significant level of staying (SLS) were fixed at 0.15.

Table 7.6 Factors determining an individual's cultural competence

Model (dependent variable)	Explanatory variables[a]	Parameter (S.E.)	*P*-value
Model 1 (competence to knowledge shared by informants of all seven Roviana villages)			
(*N*=32)	Cash income (log)	−0.015 (0.006)	0.0205
	Intercept	0.890 (0.036)	<0.0001
Model adjusted R^2	0.172		0.0205
Model 2 (competence to knowledge shared by informants of both Dunde and Olive)			
(*N*=32)	Village	−0.1001 (0.0263)	0.0008
	Duration of marriage	−0.0046 (0.0013)	0.0021
	Housing type	0.0602 (0.0260)	0.0291
	Age	0.0013 (0.0078)	0.0841
	Intercept	0.8045 (0.0468)	<0.0001
Model adjusted R^2	0.4578		0.0024
Model 3-1 (competence to knowledge shared by informants within Olive)			
(*N*=15)	Gender	0.0495 (0.0225)	0.0478
	Housing type	0.0483 (0.0275)	0.1049
	Intercept	0.7912 (0.0180)	<0.0001
Model adjusted R^2	0.4146		0.0402
Model 3-2 (competence to knowledge shared by informants within Dunde)			
(*N*=17)	N.A.[b]		

[a]Explanatory variables were selected by a stepwise method from the following variables:
(a) Individual characteristics: age (years), gender (male = 1, female = 0)
(b) Life stage of household: duration of marriage (years), consumption unit per production unit (total food energy required for household members divided by the number of laborers in the household)
(c) Modernity: household-level cash income (monthly, calculated logarithmically), proportion of income from employment/business, occupation (any member employed as a wage laborer = 1, others = 0), housing style (permanent = 1, traditional leaf = 0), household owning an outboard motor (yes = 1, no = 0), and individual level of education (education higher than primary school = 1, primary school or none = 0)
(d) Ecological characteristics: village (Dunde = 1, Olive = 0), crop yield of household (MJ), and crop yield per area of cultivated land of each household (MJ/ha)
[b]No matched model was found at significant level of entry = 0.15

ernization. For example, Olive had primary and secondary forests near the settlement and was less influenced by modernization, whereas Dunde was located on coastal flatland with a disturbed forest, was a densely populated settlement, and was a township (meaning Dunde was more town-like and urbanized when compared with Olive). Shortly after the marriage of a head, families had more competence because such households were faced with the necessity of coping with family production problems in using the forest, i.e., cultivating gardens, collecting house-building materials, and other activities for increasing household-level production. The effects of age were also included in the model but were not found to be significant; naturally, this variable was a factor for cultural competence because older villagers serve as a source of knowledge for other household heads. Living in Western-style permanent houses also had a positive effect. This was caused by the complicated social settings. People living in such houses did not necessarily have a high cash income, but they were high in social class, e.g., chiefs and elders, because the foreign logging companies provided these people with permanent houses. These villagers were usually the sources of knowledge, and, on other occasions, they had access to information because a variety of people gathered in their homes. Interestingly, education was not a significant factor. This suggests that new knowledge came not only from schools but also from other sources or through other social networks.

7.5 Transformation of Knowledge

This chapter examined (1) whether Roviana communities shared similar ethnobotanical knowledge; (2) whether a small but clear variation in knowledge existed among villages, in addition to the shared knowledge; (3) whether the variation was explained by modernity; and (4) whether variations in the knowledge of individuals could be explained by the characteristics of those individuals. Cultural consensus analysis suggested that a body of shared knowledge existed in the villages. Therefore, in this discussion, the first attempts were made to explain why all Roviana informants shared the ethnobotanical knowledge. First, all Roviana people had genealogical relationships (either patrilineal or matrilineal) with a population that lived under the same floral environment on a small island named Nusa Roviana. People of the Nusa Roviana had expanded from this single island to a larger area through intermarriage with those of other areas in the nineteenth century (Aswani 2000; Sheppard et al. 2000); now the Roviana people live in a broad area with similar vegetation; that is, the entire group was geographically and ecologically located within a distance of only 150 km from east to west (Fig. 7.1). Additionally, all Roviana people had been interconnected through sociopolitical relationships, marriage, recent schooling, church meetings, etc. These similarities in historical background and floral environment are thought to have contributed to the high degree of shared ethnobotanical knowledge. Note that the cultural consensus was much larger than in other societies that have been analyzed in previous studies. For example, the ratio of first to second eigenvectors had rarely exceeded ten in previous studies of ethnobotanical knowledge (Atran et al. 2002; Reyes-Garcia et al. 2007). When one considers that other studies also included useful plants in their questionnaires, as was done here, this high consensus is an important feature of the Roviana people. Further study is needed to reveal why the people of Roviana share most of a large body of ethnobotanical knowledge. This was probably caused by the short period of time that has passed after the people of Roviana had been separated from other members of their source culture. Additionally, the delayed modernization of this area probably is a second main factor responsible for this high level of consensus.

However, the results suggest that, in addition to the shared knowledge, intervillage variations also exist in ethnobotanical knowledge (Table 7.3). We examined the effect of modernity, which was measured as a first principal component; the variation in this principal component was tied to the cash economy, demographics, and contacts with Western materials. Complex intervillage differences in knowledge were correlated with modernity, in complex ways, as follows. The differences in knowledge correlated with modernity when the villages were referenced to the least-modernized village (Fig. 7.2b). Contrary to our expectations, this did not occur when villages were referenced to the village with the highest modernity score (Fig. 7.2a). Additionally, the correlation was observed only for villages referenced to less-modernized villages but not when the observation was in the opposite direction (Table 7.4). From these results, one can assume that ethnobotanical knowledge was changing in the same direction as modernization occurs, although the knowledge of the individual villagers as well as the collective village knowledge differed from one village to another, in addition to the shared part of the knowledge, before it was influenced by modernization.

This interpretation requires further discussion on two topics: (1) the cause of the variation in individual knowledge before modernization and (2) the cause of the linear effects of modernization on the knowledge of individuals. On the first point, one can reasonably assume that each village society had developed unique and indigenous ethnobotanical knowledge that was adapted to each village's social and natural environmental factors, even among the Roviana people. Although not shown in the data, I observed that the villagers often had a common name that was used for two or more very different plants by people in different villages. *Kikilapa* referred to a genus of epiphyte orchids in Olive while the same name, *kikilapa*, was applied to the balsa tree (*Ochroma pyramidale*) in Dunde. Additionally, epiphyte orchids were generally called *ubutu* in Dunde while the people of Olive used *ubutu* to refer to a thick mat of moss growing on the ground. The social factors that created this variation may include migration and communication from the neighboring linguistic populations while the fact one name was used for two different things may be a result of natural environmental factors including the plant species found abundantly in the forest surrounding each village. Previous studies have suggested that intermarriage between different linguistic populations has existed since the nineteenth century (Aswani 1999; Sheppard et al. 2000). Those studies also suggest that the availability of each plant species in each living area differed slightly from one village to another, even within the same vegetation type; this was a result of the distance of each village from various habitats, such as mountains, coasts, or rivers, and/or the land use of earlier generations.

For the second point, the other analyses of this study provided suggestions on the correlative effects of modernity on changes in ethnobotanical knowledge. The villages that had access to commercial markets were thought to have developed knowledge related to the cash values of plants, whereas in remote villages such knowledge was less valuable. This finding in this study was caused by a significant correlation that was observed between modernity and the number of plants known to be useful in producing a cash income (Table 7.5). However, this same table shows that the modernity score was not affected by the methods in which foods, buildings, medicines, traditions, and tools were used in traditional domains in other areas. Thus, even though people had diverse ethnobotanical knowledge, local experience affected the uniqueness of that knowledge. Furthermore, only a recognition of cash income consistently affected this ethnobotanical knowledge, because cash was the only stable value analyzed here that was clearly linked to modernization.

The complicated nature of the effects of modernization on ethnobotanical knowledge was found even at the individual level. The most important finding from the multiple regression analysis was that individual- or household-level characteristics, such as gender and housing style, affected competence only in Olive (model 3-1 in Table 7.6), whereas in Dunde (model 3-2) the analysis found no factors that affected ethnobotanical knowledge. This difference between villages is thought to be related to social organization, as described below. In Olive, decisions were made communally. For example, the authors observed that the villagers usually consulted other villagers (elders and relatives) to discuss cultivating gardens, building houses, treating illnesses, etc., and decisions were made based on the communal discussions with village leaders. This communication helps to increase the competence of individuals to a certain extent, and this in turn increases the similarities of the knowledge gained by individuals within the village but not between villages. At the same time, this communication suggests that individuals do not need to learn all the necessary local knowledge themselves, because they can consult an elder for professional knowledge as needed. In other words, in Olive, where a local community of interdependent households still functioned, individuals had their own professional knowledge and consulted each other as required. In contrast, Dunde was an urbanized and more individualistic society, because individuals owned small plots of land and settlements were separated into individually owned houses, without communal ownership. In this village, villagers needed to collect all the materials they needed from their own territory by themselves and were not allowed to trespass on the land of others. The villagers were thus less likely to consult other villagers and to share knowledge with them. This situation made it necessary for villagers to possess all the knowledge required for survival, but no additional knowledge was needed. Thus, the interindividual variation in knowledge was high, but it was not related to specific individual features. Therefore, in the quantitative analyses, individual-level modernity and social variables predict individual knowledge significantly in a less-modernized village, but it is just the opposite in a modernized village.

Based on these findings, the author suggests that the change in ethnobotanical knowledge had two or more domains, one of which was strongly affected by modernity, while the other was never explained by it. Similar findings were reported by Reyes-Gracia et al. (2007), who found that theoretical ethnobotanical knowledge and practical skills in using plants were not strongly correlated and that their association with modernity variables differed from one village to another. Several studies have suggested that indispensable indigenous local knowledge is at risk of being lost (Atran et al. 2002; Balick and Cox 1996; Prance et al. 1987; Townsend 2000). However, this study suggests that modernization was directly related to an increase of introduced ethnobotanical knowledge such as knowledge related to the cash values of plants; that is, modernization was not related to the loss of indigenous botanical knowledge. This occurs because the shared ethnobotanical knowledge was gained independently of modernization.

When the findings in the previous chapters are considered, a change of knowledge among villagers may be the result in changes of subsistence activities and conservation practices and ideas. Previous studies have suggested that knowledge, or recognition of the ecosystem, was strongly correlated with land use behavior and with ecological biodiversity of the surrounding environment (Atran et al. 1999, 2002). Villages may risk the loss of diversified indigenous knowledge that had been adaptive to a specific microenvironment, and this in turn may create a risk of the loss of the villagers' direct or indirect contribution to biodiversity. Furthermore, an increase of Western knowledge may bring a more accurate level of biological knowledge to the villagers, but this new knowledge may not be adaptive to those living in the local environment. Even worse, Western knowledge may be biased in that villagers only gain knowledge related to how they can gain short-term for-profit-related benefits.

Chapter 8
Painful Lessons from Urban Ecology and Economics

Abstract This chapter predicts the future of Melanesian people and ecosystems, based on comparative research in the urban area of Roviana. Households in Munda town were economically vulnerable and at risk of encountering poverty and other hardships. For example, while households with regular jobs earned 3.6 times more money per hour than rural households, the urban households with no regular job, who were the majority, earned the same level as the rural households. The amount of food production either per hour or per hectare was small in urban households with or without regular jobs and was much lower than rural households. A health risk was prevalent in the urban people who were economically vulnerable and suffering from low food productivity. This result suggested the necessity of taking action to conserve the rural lifestyle and give it a viable future.

Keywords Production • Consumption • Efficiency • Economy • Demography • Health • Urban ecology

8.1 Urbanization in the Solomon Islands

The sizes and numbers of urban areas have been increasing globally as human population densities increase; urbanites tend to have higher than average incomes, cities gain increased anthropogenic land cover and related ecological changes, and cities exhibit high levels of energy consumption (McIntyre et al. 2008; United Nations 1999). Generally speaking, few city governments have been able to provide an increasing population with adequate land, services, and facilities as well as employment or other appropriate sources of income. As a result, the seemingly constant increase in the numbers of urban residents has generated problems concerning the health and welfare of people, with some of these problems caused by insufficient local food production and resulting in poverty (Cocklin and Keen 2000). Small countries, such as those of Melanesia, are the most vulnerable to economic and environmental change (Aswani et al. 2007; Cocklin and Keen 2000).

Previous chapters illustrated how rural people adapt by wisely using biodiversity. However, those chapters also implied that even rural societies are now facing conditions that cause them to participate in a market economy, a situation that may

T. Furusawa, *Living with Biodiversity in an Island Ecosystem*, Ethnobiology,
DOI 10.1007/978-981-287-904-2_8

have unexpected risks. The future of a rural village might include further development such as that of the urbanized village, Dunde. Since the urbanization of Solomon Islands has been slow, the lifestyle of urban people still depends on natural resources to some extent. Dunde is characterized by high population density and the people of Dunde have been exposed to a cash market economy (e.g., employment, a variety of stores) and have access to governmental services (e.g., health and education) and enjoy a basic level of infrastructure (e.g., road access, communication, water supply, and electricity). This chapter first describes the economic situation and the welfare of people in the urban village, Dunde, in comparison with the rural village of Olive, to help the reader understand the changes occurring in Roviana.

In this regard, understanding the decision-making activities of households and individuals is also important (Hviding 2003a; Moran 1990). In the present study, everyone in the region has been exposed to a cash economy in the past few decades. Therefore, they now carefully choose their lifestyles and strategies based on increasing job opportunities and the availability of cash income and consider the decreasing amount of available land by perhaps using land of a lower quality than in the past for gardens; some households remain dependent on traditional horticulture while others have begun activities that create a cash income (Furusawa and Ohtsuka 2006). Understanding the subsistence patterns at both the population and household levels is required to understand the effects of these recent socioeconomic changes from a human-ecological viewpoint (Morán 1990; Ohtsuka and Suzuki 1990).

8.2 Near Future of Socio-ecological Changes

Dunde households were divided into two groups for the analysis, based on the occupational status of the occupants. First, about 30 % of Dunde households depended on a regular salary from professional occupations (hereafter classified as "urban households with a regular job"). The remaining households were mainly engaged in subsistence activities without any regular salaried work ("urban households with no regular job"). In contrast, all Olive households were evaluated en bloc in this study because they lacked any marked difference in the occupations of the householders ("rural").[1]

Table 8.1 shows sociodemographic conditions of these three groups as described above, two groups in Dunde and one in Olive. Although the total fertility rates

[1] All 202 households in Dunde and 65 in Olive were visited by the author for interviews related to the following topics: (1) gender, date and place of birth, and occupation of all members, (2) reproductive history of all adult females, and (3) possession of modern goods (i.e., houses made with modern materials and possession of outboard motors, chainsaws, generators, and rainwater tanks). The following households were randomly selected and used for the collection of detailed data related to subsistence and dietary habits: 16 households (58 male and 53 female members in total) from Dunde and 15 households (60 male and 52 female members in total) from Olive. Five members of the town households had regular jobs, and 11 had no regular job.

Table 8.1 The sociodemographic and modernity features of the study villages[a]

	Urban: with a regular job	Urban: no regular job	Rural
No. of households (HHs)	61	141	65
(No. of participant HHs)	(5)	(11)	(15)
No. of individuals	335	730	379
(No. of participants)	(36)	(75)	(112)
Total fertility rate	4.80	4.09	5.57
Permanent houses per HH	1.00	0.80	0.28
No. of outboard motors per HH	0.44	0.18	0.11
No. of rainwater tanks per HH	1.15	0.74	0.17
No. of stores per HH	0.07		0.03

[a]Based on the same data as Table 2.2; HH indicates households

(TFRs) were lower in both the urban households with no regular job (4.09) and those with regular jobs (4.80) than in rural households (5.57), even the former were much higher than the average (3.1) for developing countries worldwide (Population Reference Bureau 2004), indicating that the population of the study area is rapidly increasing. Note that people in this region have typically had a high fertility rate (Nakazawa and Ishimori 2004) and this pattern was possibly a relatively new phenomenon as a reaction to a population decrease which started in the latter part of the nineteenth century (Bayliss-Smith 2006). However, the fertility rate has been gradually decreasing since the 1980s (Nakazawa and Ishimori 2004). The possession of modern goods (house with permanent building materials, outboard motors, and rainwater tanks) was highest in urban households with regular jobs, followed by urban households with no regular job and then rural households.

Table 8.2 shows the socio-ecological variables of the study households as well as defines the calculations used to determine the consumer units (CUs) and producer units (PUs) of each household. For this purpose, an application of the consumer unit (CU) and producer unit (PU) concepts allowed comparisons of the amounts of foods produced and money generated by the households. In this study, the CU of a household was defined as the sum of each member's energy requirements, because it was generally assumed that foods or commodities were distributed proportionately according to each individual's energy requirements (Umezaki et al. 2000). Note that the CU of an adult male with an average body weight of 62.8 kg was defined as 1.0 in this book; the needs of each individual were calculated based on the relative energy intake requirements while considering gender, age, body weight, and reproductive status (Food and Agriculture Organization/World Health Organization/United Nations University 2004).[2] The author's observations disclosed that those

[2]For example, a 3-year-old boy weighing 11.1 kg was considered as $CU=0.47$. Comparatively, $CU=0.90$ and 0.73 for a 36- or a 74-year-old female or male weighing 72.4 kg or 49.4 kg, respectively, reflecting different demands for energy. The food energy requirements of lactating or pregnant females were assumed to be 2100 kJ (500 kcal) or 850 kJ (200 kcal) higher than those of non-lactating and nonpregnant females, respectively, of the same age and body weight.

Table 8.2 Ecological and economic variables of urban and rural households (mean [SD])

	Urban: with a regular job [UW]	Urban: no regular job [UN]	Rural [R]	Tukey's $P < 0.05$
Consumer unit: CU [a]	6.0 [0.5]	5.4 [1.9]	5.8 [2.2]	
Producer unit: PU [b]	3.0 [1.4]	2.9 [1.4]	3.3 [1.5]	
CU/PU	2.3 [0.9]	1.8 [0.6]	1.8 [0.5]	
Garden area (m^2)	111.6 [153.7]	504.1 [418.4]	846.8 [510.8]	R>UW
Garden area (m^2)/CU	20.1 [29.3]	98.1 [81.4]	155.3 [72.3]	R>UW
Crop yield (MJ)/day	1.16 [1.59]	7.25 [8.82]	19.72 [12.50]	R>UN, UW
Crop yield (MJ)/day/CU	0.20 [0.28]	1.34 [1.55]	3.56 [1.82]	R>UN, UW
Net cash income (SBD (USD))[c]/day	147.1(≈20.6) [120.7(≈16.9)]	18.1(≈2.5) [20.5(≈2.9)]	12.6(≈1.8) [11.2(≈1.6)]	UW>UN, R
Net cash income (SBD (USD))/day/CU	25.1(≈3.5) [20.6(≈2.9)]	3.1(≈0.4) [2.8(≈0.4)]	2.5(≈0.4) [2.6(≈0.4)]	UW>UN, R
Cash expenditure (SBD (USD))/day	31.8(≈4.5) [15.8(≈2.2)]	16.4(≈2.3) [6.8(≈1.0)]	11.9(≈1.7) [6.1(≈0.9)]	UW>UN, R
Cash expenditure (SBD (USD))/day/CU	5.4(≈0.8) [2.8(≈0.4)]	3.3(≈0.5) [1.6(≈0.2)]	2.2(≈0.3) [1.2(≈0.2)]	UW>R

[a]An adult male with an average body weight (62.8 kg) was defined as 1.0 *CU* (consumer unit), and each individual was labeled with a relative ratio on the basis of his/her energy requirements (Food and Agriculture Organization/World Health Organization/United Nations University 2004)
[b]*PU* (producer unit) was defined as: adults 18–70 years of age = 1 and other individuals = 0
[c]The net income from retail sales was calculated based on the amounts earned and expended for necessities

over 70 years old seldom conducted horticultural and cash-earning activities, but other adults and adolescents (18 years or older) of both genders played major roles to similar extents in both horticultural and cash-earning activities. Therefore, the PU was defined as 1 for all 18–70-year-old males or females but 0 for everyone else in this analysis. There were no significant differences in consumer units (CU: 5.8–6.0 on average), producer units (PU: 2.9–3.3), or the CU/PU ratio (1.8–2.3) between groups. Nevertheless, there was a slight increase in the CU/PU ratio in the urban households with regular jobs; this reflects the fact that these households, as is their custom, provided rooms for their relatives who were born in rural villages and attended school in Munda.

To compare agricultural production, cultivation garden area and crop yield were measured for consecutive 14 days. The total garden area per CU was significantly larger in rural households (155.3 m^2 on average) than urban households with regular jobs (20.1 m^2); that of urban households with no regular job (98.1 m^2) was intermediate.[3] More importantly, the crop yield of rural households (3.56 MJ/day/CU in average) was significantly larger than that of the urban households with regular jobs (0.20 MJ/day/CU) as well as that of those with no job (1.34 MJ/day/CU).

[3]Tukey's multiple comparison test ($P < 0.05$).

For comparing cash income, the author visited all adults in participating households every evening for 28 consecutive days, asking about the sources of any cash they had obtained and about any cash they had spent. The net income from retail sales and catering (e.g., from a bakery) was calculated, based on the amounts earned and expended for necessities (e.g., wheat flour). As expected, the cash income of urban households with regular jobs was higher than that of urban households with no regular job and of rural households with no regular job (25.1 Solomon Island dollars (SBD) (or ≈ 3.5 USD)/day/CU, 3.1 SBD (or ≈ 0.4 USD)/day/CU, and 2.5 SBD (or ≈ 0.4 USD)/day/CU, respectively). The cash expenditure per CU by the urban households with no regular job was not significantly different from those with regular jobs. This occurred even though the average net income (18.1 SBD or ≈ 2.5 USD)/day) was slightly higher than the expenditure (16.4 SBD or ≈ 2.3USD)/day); the expenditure per CU (3.3 SBD or ≈ 0.5 USD)/day) exceeded the income per CU (3.1 SBD or ≈ 0.4 USD) /day) in the former. Rural and urban households with regular jobs naturally had low cash income and low crop production, respectively, but the fact that the urban households with no regular job had both a low cash income and a low crop production suggested the existence of financial hardships in these households.

Next, the cycle of shifting cultivation was analyzed. Table 8.3 shows the duration of cultivation and fallow for horticultural land[4]; urban households with regular jobs were not included in the analysis because few households conducted horticultural activities. The duration of the fallow period was significantly shorter in urban households (3 years on average) than in rural ones (13 years), despite the fact that the duration of cultivation in the same garden was similar between those two types of households (5–6 years for gardens abandoned within the 1 year). Note that rural Olive people had gardens not only on the main island of New Georgia but also on the barrier island, where crops grow for longer cultivation periods with a short fallow period with the rich soil nutrients available there (Chap. 4). Therefore, the difference between urban and rural households was larger when the durations of the growth of gardens in only New Georgia were compared with each other.[5]

Further analyses were made to determine the efficiency of cultivation in generating cash-earning activities based on production per time or land area. Table 8.4 shows the efficiencies of cash earnings in marketing and food production in horticulture.[6] While households with regular jobs earned 3.6 times more money per hour

---

[4] In either Dunde or Olive, household heads were interviewed about the duration (in years) of the fallow and cultivation periods for all gardens under use and abandoned in the past year. To calculate land productivity, all crops harvested from the gardens were weighed every day for 14 consecutive days. The analysis of Olive (Chap. 4) explains the details.

[5] For the gardens in Olive on New Georgia Island, the length of cultivation of gardens under use, that of gardens abandoned within the last year, and the length of the fallow period were 1.5 ± 1.0 (N = 22), 1.8 ± 1.5 (19), and 26.5 ± 33.2 years (15), respectively.

[6] Time allocation studies were conducted, based on modified spot-check observations for individuals, for all adult members aged 18 years or older in the participant households (eight males and nine females in urban households with regular jobs, 18 males and 18 females in those with no regular job, and 26 males and 25 females in rural households). The households were visited every

Table 8.3 The durations of cultivation and fallow for all gardens made by 11 urban households and 15 rural households (mean [SD])

	Urban: with no regular job	Rural	Wilcoxon's *P*
Length of cultivation (years)			
Gardens under use (*N*)	1.6 [2.3] (18)	2.6 [4.4] (32)	NS
Gardens abandoned within last 1 year (*N*)	6.3 [7.9] (5)	5.6 [12.7] (22)	NS
Length of fallow (years)[a]			
Gardens under use and abandoned (*N*)	3.0 [2.2] (19)	13.4 [12.5] (28)	0.0009

[a]Gardens made immediately after the clearance of primary forest were excluded from the calculations of the fallow period

Table 8.4 Efficiencies of cash earning and food production

	Urban: with a regular job	Urban: no regular job	Rural
Time spent in horticulture per person per day (h)[a]	0.15	0.73	1.10
Time spent in cash-earning activities per person day (h)[a]	3.68	1.55	1.07
Net income (SBD(USD)) /time spent in cash-earning activities (h)	12.04 (≈1.69)	3.58 (≈0.50)	3.31 (≈0.46)
Food production from gardens (MJ) /time spent in horticulture (h)	2.26	3.03	5.28
Food production from gardens (MJ) / garden area (ha)	104.0	143.9	232.9

[a]Based on the spot-check observations of 17 adults in urban households with regular jobs, 36 adults in urban households with no regular job, and 51 adults in rural

(12.04 SBD or ≈ 1.69 USD) than rural households (3.31 SBD or ≈ 0.46 USD), the urban households with no regular job earned the same level (3.58 SBD or ≈ 0.50 USD) as the rural households. The amount of food production either per hour or per hectare was smaller in urban households with or without regular jobs and was lower than rural households. Urban households with no regular job spent much less time engaged in cash-earning activities than those with regular jobs and slightly more time than rural households.

90 min from 7:00 to 20:30 in Dunde and every 80 min from 7:00 to 20:20 in Olive, respectively, as discussed in Chap. 4. This survey provided 1071, 2268, and 3570 spot-check observations for town households with regular jobs, those with no regular job, and rural households, respectively. The activities of the participants were recorded in detail, and they were categorized into horticultural, cash earning, and "others" in the analysis.

Finally, the diversity of economic activities among households was analyzed. Figure 8.1 shows each household's net cash income per CU. Clearly, large inter-household variations existed in all groups, but the overall cash incomes were much higher in urban households with regular jobs than in the other two groups. Almost all income of the urban households with regular jobs came from their occupations, including employment (i.e., public officer, nurse, or mechanic) and businesses (i.e., retail stores); two households running stores gained especially large amounts of income. In urban households with no regular job, selling marine resources provided two households with the largest amount of money, while wage labor ("employment/labor") and retail sales, including catering and baking, accounted for the bulk of the income among four and three households, respectively. In rural Olive village, 14 out of 15 households earned their income by selling marine resources (exclusively *Nassarius* shells); this activity contributed to 55.9 % of the total income on average, ranging from 9.2 to 100 % among the households. Three households had income from wage labor within the customary land. Additionally, two households depended largely on remittance from the relatives, who were engaged in manual labor for the logging company, which was operating in remote areas some distance from the village.

Note that the net cash income was not correlated with the amount of crop yield within each group nor in all households in a pool. This fact suggested that the low crop production was not necessarily supplemented by cash income. Overall, the majority of households (i.e., households with no regular job) in the urban area suffered from both low food productivity and from having a low cash income.

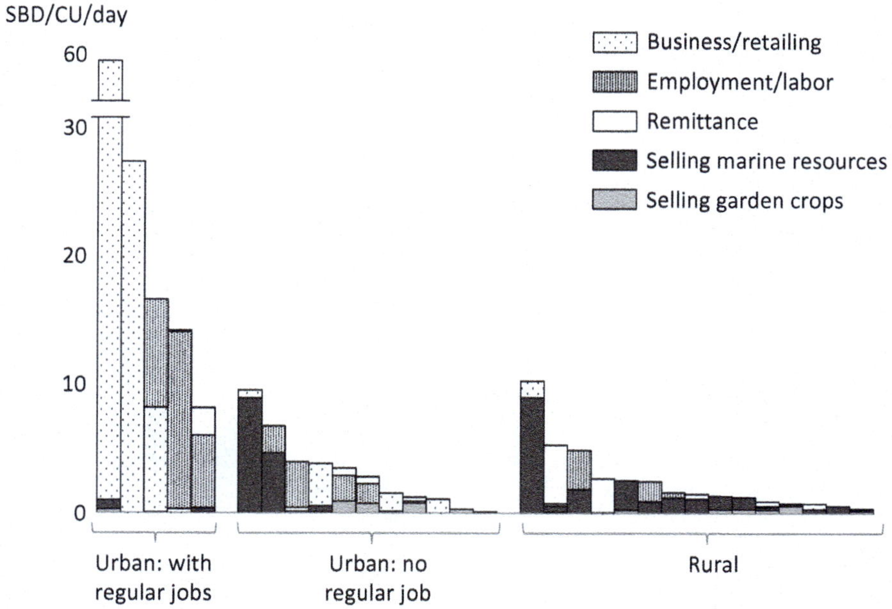

Fig. 8.1 Income of each household per CU by source

Figure 8.2 shows expenditures per CU for each group. Urban households (both with and without regular jobs) typically spent substantial amounts of money to purchase local crops and fishes, while rural households rarely purchased them. Expenditure for imported or commercial foods (e.g., imported rice, flour, tinned fish) was the highest for the urban households with a regular job (87.9 SBD/CU or ≈ 12.3 USD for 28 days), followed by the urban households with no regular job (62.2 SBD or ≈ 8.7 USD) and the rural households (35.7 SBD or ≈ 5.0 USD). These results suggest that the urban people needed to spend more money to obtain food, while the rural people can produce foods in their own territory. Urban households, even those with no regular job, spent more money for fuel than the rural households because the urban people had less access to firewood in their local areas.

These results stand in contrast to a previous study in Port Moresby, the capital of Papua New Guinea (Umezaki and Ohtsuka 2003). They reported that cash earnings of households relying on informal sector activities (e.g., selling betel nuts, scones, or beer) were similar to or higher than cash earnings of those in the formal sector (e.g., public servants, drivers, and carpenters). This pattern was observed not only in the net income but also in the income per labor hour or that per CU. Port Moresby is a large city with a population of 248,948 people in 2000 and had expanded and improved its international cash economic markets based on the production of oil and natural gas. In contrast, Munda is a small urban center consisting mainly of customary lands. Even though it may be possible for the cash economic conditions to become as good as those in Port Moresby in the future, the people without regular jobs will continue to suffer from the double burdens of a low crop yield and low

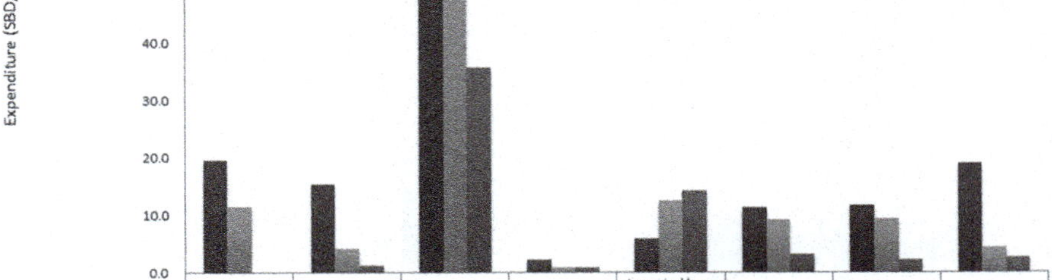

	Local crops	Local fishes	Imported/commercial foods	Local luxury items (betel)	Imported/commercial luxury items (tobacco, alcohol, tea, etc)	Fuel	Daily necessities (soap, mathces, basket, etc)	Others (transport, education, medical, etc)
Urban: with regular job	19.6	15.3	84.9	2.2	5.9	11.3	11.6	18.9
Urban: with no regular job	11.5	4.2	62.2	0.8	12.4	9.2	9.3	4.4
Rural	0.2	1.3	35.7	1.0	14.1	3.1	2.3	2.7

Fig. 8.2 Expenditure of each group per CU (for 28 days)

cash income, as shown in this study. Additionally, population growth and an increased desire for monetary prosperity are already manifesting themselves in the non-urban areas in Makira Province of the Solomon Islands (Fazey et al. 2011). Because urbanization has expanded in and near many customary lands throughout the Solomon Islands, one can reasonably assume that the people are highly vulnerable to urbanization.

8.3 Health and Survival

Further analyses addressed health and survival. Although insufficient food consumption and relevant diseases (e.g., infectious diseases) seem to be important health and even survival-related problems, previous studies, including my own, have recently found low incidences of malaria, other infectious diseases, and malnutrition in the country (Eason et al. 1987; Furusawa et al. 2011a, b; Yamauchi 2007).[7] Anthropometric measurements[8] revealed that 40 and 33 % of adult members of the study households were overweight (BMI≥25.0 kg/m^2) in urban (Dunde: N=55) and rural (Olive: N=45) areas, respectively. Note that malnutrition (BMI<18.5 kg/m^2) was very rare in both locations, with five individuals in urban and one in rural households. Table 8.5 shows the association of the prevalence of overweight people with crop yield per CU and cash income per CU for the urban and the rural people, who were grouped into tertiles, based on the household-level crop yield and cash income per CU; in this analysis, the urban households with and without regular jobs were pooled. This result showed that being overweight was prevalent in not only high-cash-income households but also in medium- and low-cash-income households. The condition of being overweight was more prevalent in low-crop-yield households in the rural village than in the other locations. Although the small number of adults examined limits the interpretation of these results, these results suggest that households with a low economic status were also at risk of noncommunicable health problems, such as obesity.

A previous study also suggested that rural residents had sufficient local food production (garden crops and fishes), while urbanized residents did not (Furusawa et al. 2011a). Interestingly, the prevalence of noncommunicable diseases was low in the former, suggesting that lifestyles with sufficient local production, subsistence, and physical activities could have decreased the incidence of noncommunicable disease risk. A previous finding partly supported the suggestion that the abundance

[7] Health statistics have suggested that the prevalence of malaria in Western Province had decreased from a peak of 506 persons with malaria per 1000 population annually in 1994 to 58 patients per 1000 population annually in 2005; furthermore, a mass administrative survey was conducted soon after a disaster in 2007, with only 78.5 in 1000 population patients found to be positive for malaria (Malaria Department, National Gizo Hospital, personal communication).

[8] The height and weight of all of the participants were measured, to the nearest 1.0 mm using a field anthropometer (TTM, Japan) and to the nearest 0.1 kg using a portable digital scale (Tanita model 1597, Japan), respectively.

Table 8.5 Prevalence of overweight adult members (≥18 years old) in urban and rural households. Households were classified into high, medium, and low tertiles[a]

	Urban (Dunde)			Rural (Olive)		
	Criteria	*N*	% overweight	Criteria	*N*	% overweight
Cash income/CU tertile	SBD/day			SBD/day		
High	264–1675	19	32	71–284	10	40
Medium	78–226	22	45	35–69	21	33
Low	3–43	13	46	10–26	8	29
Crop yield/CU tertile	MJ/day			MJ/day		
High	1.0–5.0	19	47	4.6–7.1	18	22
Medium	0.1–0.6	17	29	2.7–3.9	13	23
Low	0	18	44	0.6–2.6	14	57

[a]Only willing volunteers were studied, so not all members of the study households were measured. As an indicator of nutritional status, the body mass index (BMI: kg/m^2) was calculated; a BMI of 25 or greater was classified as overweight (WHO 2000)

and availability of natural resources, e.g., fishes, were directly related with health status in this area (Aswani and Furusawa 2007). In summary, this comparison showed that obesity and overweight conditions were prevalent in the urban people who were economically vulnerable and suffering from low food productivity. In this regard, this reflects a global tendency; when commercial foods become inexpensive and available for either low-income countries or households, then obesity rates tend to increase (Drewnowski and Popkin 1997). More importantly and specifically for Melanesian countries, these imported foods were less expensive than local crops in terms of energy per unit price. For example, rice and flour contain 3.4 MJ and 5.7 MJ per 1 SBD (≈0.14 USD), respectively, while sweet potato and cassava have 2.2 MJ and 5.2 MJ, respectively, based on the prices sold in stores in both study sites and in the Munda Town Market (the author's unpublished data). The energy per unit price was higher for cooking oils and sugar. Thus, the households with inadequate horticultural production tended to depend on imported high-energy foods. Except for the high-income households that can diversify their food choices, urban residents face these conditions that affect their quality of life.

Based on these comparisons between rural and urban households, the future of the rural life should be discussed. If the rural households follow the same pattern as the urban households (i.e., population growth, integration into the market economy, and loss of customary governance), the majority of these rural households will continue to suffer from having a low cash income, low crop yields, and poor ecological services (Chap. 5), and their members will be at a higher risk of illness. Although Ohtsuka et al. (1995) pointed out that an increase in cash income without large inter-household differences was one effective strategy for coping with the increased population pressure in rural Papua New Guinea, this study suggests that such a strategy will not meet with success in the Solomon Islands. That is, an increase in cash income will not necessarily be achieved. Additionally, as Fazey et al. (2011) pointed out, communities are becoming more vulnerable to the future impacts of population growth, the desire for monetary prosperity, and factors actively changing in the global environment (e.g., climate change and an increase in fuel prices). Because the people were powerless to improve the poor economic market and to deal with overpopulation effectively, it will be necessary to find solutions by improving the use of resources supplied by their land.

Chapter 9
Customary Rules and Wisdom Related to Conservation

Abstract This chapter provides qualitative information such as customary rules related to forest use and ecosystem management with the goal of exploring the possibility of integrating the ways that local people modify forests into new biodiversity conservation programs. Customary rules related to forest conservation guide the actions of villagers, such as rules related to the forest reserve, the use of white beech (*G. moluccana*)—a tree used to make canoes—and sago palm leaves (*Metroxylon* spp.), an essential roof and wall material in traditional houses. These decisions suggest that local leadership, especially the leadership of the chief and church leaders combined with traditional ecological knowledge of local people, plays an important role in conservation. However, the recent increase in the size of the human population and commodification of resources has caused the people of Roviana to face the risk of overexploitation of natural resources and to drive them to integrate Western active conservation programs into their resource management techniques.

Keywords Customary resource management • Traditional conservation ethics • Chief • Social institution • Compensation • Resource protection • Folklore

9.1 Social and Cultural Aspects of Conservation

Previous chapters provided quantitative evidence related to the diversity of the interactions between the Roviana people and the natural world; in contrast, this chapter focuses mainly on qualitative aspects such as changes of traditional and customary rules related to conservation and ecological wellness. Throughout this book, the rural people are described as having contributed to the creation and conservation of biodiversity while they intended to enhance their livelihoods and to enjoy the benefits of the ecosystem services in their territories. Such practices lead to the conservation of various forest types and thus the preservation of the diversity of tree species; however, the findings should be interpreted carefully where their practices were specifically intended to conserve and create biodiversity, because such

T. Furusawa, *Living with Biodiversity in an Island Ecosystem*, Ethnobiology,
DOI 10.1007/978-981-287-904-2_9

conservation might have been a by-product of the human activities that were actually designed to create short-term benefits (Smith and Wishnie 2000; Foale et al. 2011).

9.2 Biodiversity Conservation as a Social Institution

Customary rules related to forest conservation guide the actions of villagers in Olive, such as rules related to the use of white beech (*G. moluccana*), a tree used to make canoes (Fig. 9.1). When a person needs a new canoe, he/she must find a young white beech tree and mark it to inform other residents of its intended use. Then, when the time comes to make the canoe, he/she must ask for the customary chief's permission. White beech typically grows in natural forest gaps (Bayliss-Smith et al. 2003; Whitmore 1966); therefore, human forest modification, such as the small-scale removal of trees in reserved forest (Chaps. 3 and 5), contributes to the sustainable use of this resource. Another rule also provides guidelines on the gathering of sago palm leaves (*Metroxylon* spp.), an essential roof and wall material in traditional houses; the trunk must be left untouched, and only the leaves can be removed because a sago palm takes a very long time to regrow once the trunk is cut. Four leaves should also remain on the tree, because it will die if all the leaves are removed (Fig. 9.2). If a villager violates this rule, he/she will be penalized by the chief (e.g., compensation payment). In fact, the author observed these types of occasions. For

Fig. 9.1 Building a canoe from white beech (*qoliti*)

Fig. 9.2 Sago palm (*edeve*) after collection of leaves for roofing material for a house: four leaves were left

example, a dispute occurred when all villagers were engaged in the collection of sago palm leaves for repairing the school building as a group activity for the community workday. A group of women claimed that some other villagers selfishly cut all leaves of sago growing near the settlement, rather than leaving the required four leaves, to make their quota easy, and they believed that such unscrupulous people should be reported to the paramount chief. Village leaders discussed and handled this issue by requiring compensation from those who tried to avoid work in this way.

Chapters 3 and 5 disclosed the important roles that the "reserve" forest plays related to biocultural diversity. According to the interviewees, the rural people established the reserve forest in the 1970s–1980s based on the recommendation of the Christian Fellowship Church leader and the local chief. The reserve was established when the people intended to expand a coconut plantation for cash income and when a logging operation had also been active near their territory. These decisions suggest that local leadership, especially the leadership of the chief and church leaders combined with traditional ecological knowledge of local people, plays an important role in conservation.

Another factor is spiritual awe of the forest. During the fieldwork, the author experienced situations where people expressed awe and were afraid of the forest spiritually. For example, the author was required to not show disrespect in the forest, such as by causing unnecessary noise while visiting it. The forest expert stated that *tomate* (ghost or spirit) or *malivi* (monster or giant) would make mischief for the people if they broke this rule. These *tomate* were frequently described as spirits of their ancestors and *malivi* as the indigenous giants on the islands. Undesirable experiences in the forest, such as centipede bites, are usually recognized as the result of poor behavior while visiting the forest. In one case, a worker from another province who was employed at the logging camp was bitten by a centipede while he was in the sacred forest (*hope*); he suffered from pain for several days and finally asked the paramount chief to pass on his apology to the spirits of the ancestors. In Roviana folklore (called *vivinei malivi*), one creature described as *malivi* lived in *soloso* (inner forest). A story titled *Karua tamatasi koreo, sa malivi meke sa kara* (literally "two brothers, a giant, and a singing parrot") was compiled in *Na Buka Vivinei Malivi pa Zinama Roviana* (Roviana Customary Stories Book), a leaflet edited by the Western Province Government (1991). The story tells of a young boy who entered the inner forest in search of fruit but encountered a giant. The boy desperately tried to escape as the giant pursued him. A wise singing parrot (*kara*) began to speak, sounding exactly like the boy, distracting the giant, and allowing the boy to escape. Today, all Roviana people live as professed Christians. They always prayed to God in the Christian style before entering the forest; they did the same before boarding their boats and going out to sea. Even though they prayed to the Christian God for safety, they believe that a traditional power (e.g., chief) is needed to resolve problems in the forest or problems with the spirits of the ancestors, as observed in the case of the logging worker.

9.3 Practices and Institutions Under Socioeconomic Changes

Some measure of biodiversity has already been lost in the urban village of Dunde, primarily because the reserve and mangrove forests no longer exist or have been severely degraded or diminished in size (Chap. 5). Consequently, the urban people did not enjoy the ecological services of useful plants that would otherwise be expected in those habitats. For example, the mangrove has been an important useful plant in this area, but few people used mangrove in the urban areas where mangroves were rare. This loss of biodiversity causes the people of Dunde to purchase other tree resources with cash or to use modern materials to compensate for this loss (Chap. 8).

In recent years, under an expanded market economy, widespread commercial logging has been conducted, and industrial reforestation is now being implemented in logged areas. This trend, if it continues, could destroy the residents' source of livelihood, triggering a decrease in the number and variety of forest species. Although industrial timber plantations will contribute to increased biomass in the forests, plantations may interrupt the original cycle of biodiversity and disrupt the relationships between humans and forests. People will lose access to vital ecosystem services provided by the current range of forest types in the area if logging and the planting of cash crops continue unabated (Fazey et al. 2011; Rennie 1991). The development of timber plantations will not replace the ecosystem benefits currently available from forests that have been managed using traditional methods. Previous studies have suggested that shifts from subsistence agriculture to cash crops have adversely affected the availability for work for local people, their economic status, and natural productivity in other parts of the Solomon Islands (Bennett 2000; Garonna et al. 2009; Hviding 1996; Hviding and Bayliss-Smith 2000; Kenter et al. 2011). Additionally, a correlation may exist between this loss of biodiversity and the loss of traditional botanical knowledge because socioeconomic changes affect both biodiversity and traditional knowledge (Chap. 7).

Furthermore, changes in socioeconomic conditions also influenced how people think about conservation. For example, at the outset of a logging operation, the companies usually employ local people to identify customary sacred places (*tabu ples* in pidgin and *hope* in Roviana) allowing timber companies to avoid disturbing such sites; national forestry policy and agreements with local people directed this action. According to several village informants, this activity played a role of establishment and local protection of sacred forests (Fig. 9.3).

Recently, the Olive village leaders were observed discussing a resource-related concern. They made a rule that if a villager wants to cut *Calophyllum* spp. and vitex (*Vitex cofassus*) trees, then the villager needs to get permission from the village leaders first. The local people use these two species most frequently (Chap. 5), while loggers also primarily target these two species (Chap. 6). The informants

Fig. 9.3 A tree painted with the word *TABU* indicating that this area is a *tabu ples* (pidgin term for the sacred place) and that people should refrain from logging this site

stated that they created this rule based on advice from governmental forestry authorities. According to the informants, pressure from both the logging company and the villagers increased on these trees. The villagers could not cut these huge trees until they started to use chainsaws. The use of large trees as building materials had occurred rather recently because these species required the use of saw milling machines to saw the timber into boards. The use of these trees increased rapidly because the sawn boards made beds that were more comfortable for the villagers than an uneven layer of palm trunks. That is, the changed lifestyle and new technology created pressure on these two species. Compliance to the new rule restricting the cutting of *Calophyllum* spp. and vitex was low compared with compliance with traditional rules. This fact also suggests that even the villagers' recognition and actions apparently had adverse effects on important species which were newly protected; this observation was biased by Western views to judge that such actions and recognitions evidenced the lack of traditional conservation ethics because they continued to conserve some species and vegetation following locally practiced ethics.

Another conservation case in Roviana shows how active conservation is ongoing through collaboration between local communities and outside professionals. A scholar, based on his long-term field observation of the behavior of local villages, traditional knowledge, and customary institutions, established several community-based marine protected areas (MPAs) near villages of Roviana. The MPA project received financial support from the MacArthur and Packard Foundations as well as the National Science Foundation and Conservation International (Aswani and Furusawa 2007; Aswani and Hamilton 2004; Aswani et al. 2007; Aswani and Lauer 2006). The protocol of MPA establishment calls for integrating marine and social science research with indigenous knowledge. Since 2008, this scholar, named Aswani, and his local and international collaborators have established 28 MPAs, which cover approximately 5000 ha of diverse marine habitats (Fig. 9.4); the majority of them serve as "no-take" MPAs that are permanently closed to all resource-harvesting activities. He established these MPAs based on observations of the behavior of local villagers during fishing and gathering activities as well as on marine ecological surveys; he also paid careful attention to customary sea tenure to improve local participation and compliance in the MPA efforts (Aswani and Hamilton 2004; Aswani and Lauer 2006; Mills et al. (2013)). Throughout these projects, he noted that strong customary management (i.e., high compliance to protection guidelines) benefitted villagers. The villagers were able to catch larger fish near the MPAs because the protected ecosystems allowed increased fish reproduction and growth (Aswani et al. 2007). Additionally, based on the consensus of village leaders and members, a village was even observed to independently propose expansion of the MPAs. The villagers said that, based on their knowledge and experience, the larger MPA would be more effective in protecting their resources.

Previous studies focusing on the resource management program in Melanesia have suggested that human population densities had been too low to generate pressure on resources until recently and, therefore, the conservation ethics did not evolve because they were not needed (Foale et al. 2011). Customary resource tenure and management had been developed to manage relationships between different popula-

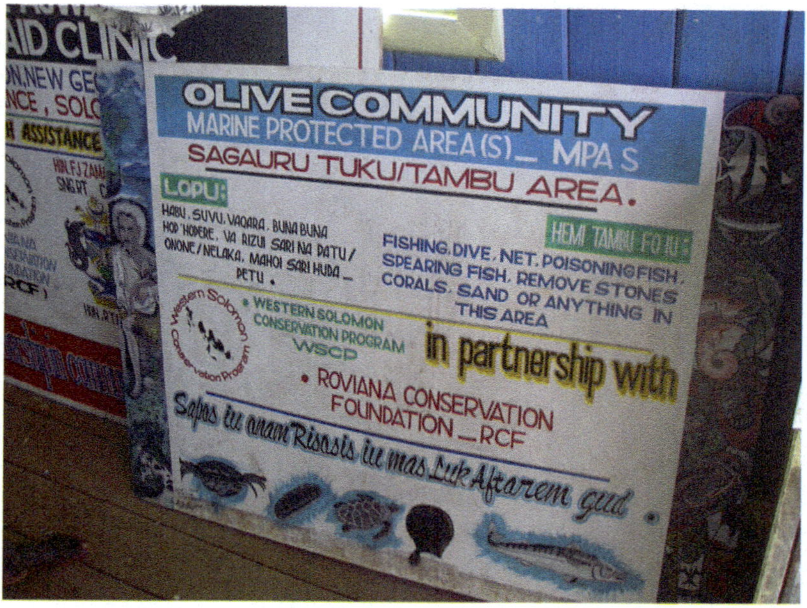

Fig. 9.4 A board describing the marine protected area (MPA) near Olive village

tion groups; this was done rather than for the protection of resources or food sources themselves (ibid.). A similar observation has been made in other societies (Smith and Wishnie 2000). However, this study found management rules within the community, and such management techniques were effective in sustaining rare species and in making them available for people. Therefore, it is reasonable to determine that the people of Roviana traditionally had a form of active conservation practices or at least have measures that were designed to sustain the availability of resources within their territories, even though many of the conservation efforts might be a by-product of subsistence activities. However, the recent increase in the size of the human population and commodification of resources has caused the people of Roviana to face the risk of overexploitation of natural resources and to drive them to integrate Western conservation ethics into their resource management techniques.

Chapter 10
Living with Internal and External Diversity

Abstract This chapter summarizes the findings, suggests ways to harmonize the interactions of humans and the natural world of Melanesia, and proposes how this system sustainably benefits the people of Roviana. Successful strategies for harmonizing community welfare and biodiversity will probably include (1) the use of diversified strategies to allow villagers to continue to receive benefits from ecosystem services, (2) the strategy that balances both low-risk and risk-prone methods of earning income, (3) conservation of biodiversity and land in a method that supports these strategies, (4) building and continuing consensus related to the sustainable management and wise use of the forest, and (5) integrating traditional leadership and knowledge for achieving these conditions. Additionally, external support is needed to empower the community with internal diversity toward harmonized interaction between humans and the natural world. Responses to external negative impacts on the island's resources may also cause the villagers to think about how they can manage the land sustainably. Recent global forest conservation initiatives can play important roles, only when they are conducted in accordance with the internal diversity. Lessons from the Solomon Islands will be useful for other societies.

Keywords Risk averse • Risk prone • Internal diversity • External factors • Harmonization • Consensus formation • Future of biocultural diversity

10.1 Living with Biodiversity

This final chapter is devoted to discussing the future of living with biodiversity based on the findings from the transdisciplinary approach used in this study of the Solomon Islands where rich biological and cultural diversity has been sustained but is now at a risk of dramatic change. This discussion addresses the following questions. What kinds of factors have determined sustainability versus a loss of biocultural diversity? How can the people of Solomon Islands recognize the natural world and the need for the conservation of biodiversity using traditional methods or under socioeconomic change? How can the society of Solomon Islands achieve that

T. Furusawa, *Living with Biodiversity in an Island Ecosystem*, Ethnobiology,
DOI 10.1007/978-981-287-904-2_10

harmony under changing socioeconomic conditions? The implications of the answers to these questions for other societies are also discussed.

10.2 Risk-Averse and Risk-Prone Subsistence Strategy

The first discussion here addresses lessons learned from the rural people's strategy for integrating activities related to a market economy and balancing them with traditional subsistence agriculture and biodiversity. The findings (especially Chap. 4) suggest that the productivity per shifting cycle of cultivation was highest on the barrier island among the various geographical locations analyzed here. This cycle was supported by the natural characteristics of the land (soil rich in phosphorus) and the social system (communal use of Raqata) of the island. Additionally, the villagers found the main island more useful for conducting new economic activities such as planting perennial cash crops rather than for horticulture. Planting such cash crops also proved risky for the villagers, primarily, because the trees they planted as cash crops would occupy the land for a long time and because it was not clear whether such crops would provide them with enough cash income to serve as a substitute for gardening in the future. Thus, the different land uses on the barrier and main islands allowed the villagers to integrate the new cash cropping system as part of their income without endangering all their food sources; they could continue receiving the benefits of sustainable production from the barrier island. Moreover, if they failed to invest time in cultivating on the main island, they were able to return to the communal location. Thus, the land use and shifting pattern of the Olive villagers constituted a diversified strategy for harmonizing the basic needs of the people while increasing their quality of life under the changing socioeconomic and demographic conditions.

A moral economy (i.e., a socially responsible and ecologically sustainable economy) involves the understanding that economic behavior provides more than simply economic benefits, i.e., political economy; a moral economy also includes moral and/or cultural aspects of an economy. Societies are likely to place the highest value on having a secure food source; this allows the villagers to avoid the risks associated with importing new agricultural schemes, ensures their basic daily needs are met, and allows them to avoid economic collapse (Scott 1976, 1985). Even if local societies were inclined toward accepting the new methods of supporting themselves and endeavored to increase their cash income under the changing socioeconomic conditions, one can reasonably assume that they will continue to be risk averse. A local strategy of risk aversion in several societies has been demonstrated in cases of various subsistence systems or domesticated species (Fratkin 1998); recent studies indicate that a society's collaborative efforts, such as resource control for avoiding overexploitation, also contribute to minimizing environmental risks (Feeny et al. 1990; Suga 2005). In the case presented here, risk was averted by having (1) diversified land use where different islands were used mainly for growing annual plants as subsistence crops and participating in cash economic activities, and (2) the use of collaborative control over Raqata as a communal place to ensure food production that met the collective community needs.

As has previously happened in many other Melanesian societies, the land tenure of Roviana has shifted from the traditional communal type to the use of private or household estates (Larmour et al. 1979; Schneider 1998). This shift potentially leads to limited availability of lands, including lands available for fallow, at the household level and inequality of benefits from lands within the community. Although such a shift had been slow in the Saikile clan, the land tenure on the main island was divided according to the levels of the sub-clans. For example, court records suggested that the settlement and gardens, which were located in and around Olive village in New Georgia Island, belonged to sub-clans of Saikile (e.g., The Gizo/Kolombangara Local Court, Civil Case No. 25 of 1980, and High Court of the Solomon Islands Civil Case No. 120 of 1994). However, Raqata remained under communal management; to our knowledge, no case had been raised on Raqata, though there was a case on another part of Ndora Island (High Court of the Solomon Islands Civil Case No. 304 of 1997). The villagers insisted that the Raqata area "must" be used communally and managed by the entire Saikile clan in case someone encountered a shortage of land or food. The members of this clan were also not allowed to establish boundaries or drive anyone away from Raqata as long as he/she belonged to the Saikile clan.

Why and how do the villagers follow their current land use and shifting cycles? One can reasonably assume that the villager's extensive experience in the study area has given them a deep understanding of the ecological differences between the barrier and main islands, in terms of indigenous ecological knowledge. However, as Chap. 6 implies, even if people recognized the ecological importance of particular resources, they were likely to place greater emphasis on economic benefits available from those resources. Actually, in one case, a Saikile clan member planted a timber species and cocoa in Ndora Island, although it was outside the communal land of Raqata; Ndora Island itself was potentially a target of cash economic development, and the perennial crops were expected to grow well on its soil. Therefore, the main reason Raqata continues to use communal land management and subsistence could be that the decisions are made by the society as a whole and by consensus rather than environmental determinisms. In addition, it is important to note that the decision was made not to seek individual's short-term benefits but to maintain communal benefits.

Consensus was a key factor driving and sustaining communal land use, primarily because the social control of the paramount chief of the Saikile clan continued to function (Chap. 9). In the Solomon Islands, customary leadership had gradually weakened and been displaced by religious or central government leadership (Ishimori 2007; Sekine 2001; White and Lindstrom 1997); however, in rural areas, such as Saikile, traditional leaders still play important roles in developing consensus (Ohtsuka 2004). Note that the church, i.e., CFC, played important roles in consensus formation in Roviana villages (Fig. 10.1), because all people belonged to one sect in the rural area, while the people were separated into several sects in the urban community (Aswani 2002; Aswani et al. 2007; Nakazawa et al. 2002; Foale 2001).

Fig. 10.1 A large ceremony of the CFC (Christian Fellowship Church) held in Mandou, 2006; followers from Roviana, North New Georgia, and other villagers gathered under the church leadership

Moreover, in relation to risk aversion, all rural villagers knew that the introduced economic schemes, such as cash cropping, were economically risky because they experienced a large fluctuation in the returns from copra, their first cash crop, as well as cocoa, and cash cropping efforts in the past had had little success. Governmental leaders had also informed the people that the newly introduced timber species would require more than one decade to mature from planting to harvest. Thus, the villagers who shared the risks of the introduced species as well as a high level of subsistence productivity on the barrier island were thought to be able to form collaborative relationships.

Several villagers from Dunde (urban people in Chaps. 5 and 8), when confronted with high population density and a cash economy, insisted that they needed a communal garden area on New Georgia Island. This occurred because they did not have rights on the small barrier island, but had already modified their land management to privately based subsistence or lands for timber species plantation. According to them, the change in Dunde was induced by the degraded productivity of horticulture under the intensive cultivation cycle; this increased the need for cash income and the Westernization of land ownership. These lessons from the rural and the urban communities suggested that if the villagers were to mitigate negative impacts of modernization, they needed to recognize the importance of having a combined strategy.

Of course, one would not feel optimistic when thinking about the future of the Olive villagers if their population continues to increase in the future, because the carrying capacity of the communal land also has its limit. In other words, new land would need to be cleared to feed the increasing population, and economic benefits would need to be secured from new sources of improved cash-generating economic activities on the main island. However, this possibility also creates the risk of creating the combined effects of a collapse of the sustainable use of the communal land and the failure of the new economic activities to generate an adequate cash flow. Estimating how the population would change exactly in the future would be very difficult, but the growth rate seems to have already started to slow down very gradually. For example, Nakazawa and Ishimori (2004) reported that complete parity decreased in generations born after the 1980s in a study conducted in North New Georgia Island, a site neighboring this study area. Such a strategy would use both risk-averse and risk-prone methods combined with a stabilized population size, to adapt to diverse landscapes, along with the continued use of local governance related to land use and management as well as community consensus-based management.

10.3 Valuing Biodiversity and Ecological Services

Another key finding was that villagers applied the local economic strategy to the use of both vegetation and landscape diversity, and this strategy was adapted to changing sociodemographic conditions. Note that this type of biodiversity has been conserved or created through the activities of the people with or without their intention. Chapter 3 showed that each forest class represented a different vegetative community, with relatively little similarity between communities. Assuming that primary forest represented areas with minimal human impacts, several unique species were found in human-modified forests (i.e., reserve forests and secondary forests). This suggested the possibility that local human society is able to develop a harmonized method of rural development that exhibits concern for community welfare while allowing for environmental conservation, although more time is necessary to evaluate the overall adaptability. In contrast, a recent monetary study reported that villagers were willing to pay a high proportion of their incomes to conserve ecosystem services (Fazey et al. 2011). The current study also found that materials that were not available from forests were available for purchase in the urban village (Chaps. 5 and 9).

Recent studies have suggested that few pristine natural areas remain free from any human impacts, worldwide (Kareiva et al. 2007). Regular subsistence use of the forests by humans may have resulted in multiple unique vegetative communities and higher overall biodiversity than would otherwise exist. The existence of a variety of forest types provides an essential base for the rural people's subsistence (Chap. 5). Top-down efforts to conserve pristine environments have generally failed, with a few exceptions; the focus of recent conservation efforts has shifted to a new paradigm of incorporating a productive landscape, social institutions, and human-modified forests into landscape-scale conservation efforts (Adams et al.

2004; Myers et al. 2000; Perfecto and Vandermeer 2008). Conservation activities are also more efficient if conservation forests and production lands are separated, but tropical societies (Green et al. 2005), as in this study, depend strongly on their forests for production. Conservation efforts based either on community efforts or on outsider initiatives have to allow for human uses of forests rather than trying to protect forests as pristine environments (Goldman 2003). The findings of this study suggest that conservation of virgin forest is not acceptable to the Roviana people, who live in human-modified forests; leaving forests untouched by human hands may diminish local biodiversity. Therefore, the focus of conservation must shift toward human-modified forests where the people use the natural resources in a sustainable way (Bélair et al. 2010; SATOYAMA Initiative 2010; Takeuchi 2010).

However, this human–environment relationship is easily transformed in light of ongoing socioeconomic changes. Ethnobotanical knowledge was strongly affected by modernity (Chap. 7) (Hviding 2006). Local leaders and consensus continue to play important roles in the conservation of biodiversity. However, a dilemma now exists because these leaders are also leading rural development initiatives aimed at improving the quality of people's lives (Racelis and Aswani 2011). The same situation exists at the national level. Generally, the governments of the Pacific Island countries depend on the production of natural resources for revenue. Additionally, even if a government tries to take action related to natural resource conservation (Frazer 1997), successful outcomes are difficult to achieve because most territories consist of lands with customary uses designated by the local people; a group of land owners may not agree to various conservation efforts but may accept development projects because they benefit local people. From the viewpoint of biodiversity conservation, the rural residents of the Solomon Islands, whose livelihoods depend heavily on farming and fishing, have no choice but to continue making a living using natural resources. Furthermore, even if development requires the use of resources, the local community still requires the group to reach a consensus.

The findings from the urban village suggested that the people of Roviana can apparently maintain their lifestyle without the ecological services provided by biodiversity (Chap. 5), although considerable effort is required to earn cash as an alternative to living off of the reserves and mangroves (Chap. 8). This model is not sustainable, because, as observed in Chap. 8, approximately two-thirds of the urban people earned only the same amount of cash as the rural residents. Additionally, logging operations in rural forests produce the materials urban villagers purchased and used for buildings, tools, and probably for fuel. Two factors were probably responsible for the low cash-earning efficiency of the urban villagers. One factor is related to the low profit from urbanized cash-earning activities (i.e., wage labor and retailing) and the other came from the limited participation in, or lowered efficiency of, a high-profit cash-earning activity in the area, i.e., the collection of marine resources. For the former, wages were very low for informal sector types of labor, although these jobs were major sources of cash income for these households. For the latter, the marine resource collection has been waning in or near the urban area because of environmental deterioration (Aswani and Sabetian 2009; Aswani 2002). Overpopulation, failed resource management, and expanded sea tenure also caused this deterioration. As a result, households with no regular job, which were the majority of households in the Munda area, were not able to gain enough money from any kind of cash-earning activities.

To the author's knowledge, several attempts have been made to develop sources of cash other than the logging in this region, such as mining and commercial agriculture (Allen and Sinclair 2013). However, mining provided only limited income because environmental, social, and political conditions were unstable. For example, foreign investments were withdrawal from the Gold Ridge Mine in 2014 (Solomon Star 2014a), and a recent court case related to mining in Choiseul also disrupted the situation (Solomon Star 2014b). The agricultural resources of the Solomon Islands have generated some global interest (e.g., *Canarium* nut, cacao), but the region's farms failed to produce enough products rapidly enough to provide a logistically feasibly market. Furthermore, because new logging operations continue to open, logging seems to be the logical alternative in the near future (Kabutaulaka 2006).

The local society encourages people to recognize the values of biodiversity and ecosystem services. Economic incentives may have to be provided if the human-modified forests are to be conserved, i.e., conserving biodiversity while using ecological services. However, a loss of traditional economic use, e.g., the fact that mangroves are now a low-priority resource and at risk of disappearing in urban areas, may lead to a loss of traditional knowledge and cultural diversity (Chap. 7).

This study defines a "forest reserve (rizevu)" as a "human-modified" forest with rich biodiversity that was created by a balance of use (removal of non-timber forest resources) and protection (prohibition of cash economic felling or clearance for horticulture). This type of reserve has a strong potential for conserving forest biodiversity in harmony with improving the quality of human life. Land tenure is a very sensitive issue in customary lands where a local clan has an exclusive right on land making a "no-take" type of conservation impossible in these forests; equally true is the fact that a "no-take" type of protection will not benefit the local people. Additionally, small-scale modification (i.e., daily use) directly or indirectly contributed to biodiversity. In conclusion, additional attention should be paid to the importance of "human-modified forest" or conservation in the sense of wise use; outside organizations should recommend that local people expand this form of conservation.

10.4 Diversity of Conservation Ethics

The diverse relationships between indigenous societies and their natural environment (Filer 2011) and those between indigenous societies and the outside world (West 2005) are important topics of discussion. For the latter, recent debates related to environmental anthropology concluded that researchers or conservationists believed that indigenous people traditionally had a sense of conservation ethics. However, this idea may cause disputes among those operating various conservation programs, because this study found the local people do not have a sense of traditional conservation ethics or voluntary actions that are specifically designed to conserve and create the biodiversity. The high value local people might attribute to biodiversity, as Western conservationists claim, it is not always shared by most local people (Smith and Wishnie 2000), and this fact may cause disputes related to various conservation projects, because conservation often takes a short-term loss as a trade-off for a long-term gain (Foale 2001). Furthermore, previous studies also pointed to the necessity of understanding local knowledge and practice and the direct and indirect effects of that local expertise on biodiversity (Cohen et al. 2014). In this regard, the fact that this study implied that local people have a lack of ethics may be biased by Western views and that the people in fact had different ethics in regard to conserving traditionally important species and vegetation than Westerners have come to expect. In other words, the concept of conservation ethics also varies within a community or even from person to person, because both communities and individuals are inclined to emphasize resource exploitation and forest clearance for short-term gain while they continue to conserve traditionally important species and recognize the complicated nature of the local ecosystem.

Note that the differences between population and individual levels also need to be considered when studying environmental anthropology or conservation ethics. Benefits at the population level, i.e., collective benefits, are not always a collection of individual benefits but also require costs at individual levels. Additionally, a collection of individual benefits does not necessarily result in a collective benefit for an entire population, as in the famous "tragedy of the commons" theory by Hardin (1968); that is, individuals are not willing to pay the cost for conservation of communal resources but are likely to use resources so as to maximize their own immediate benefits (Smith and Wishnie 2000).

This study asked, "How do people of Solomon Islands traditionally recognize the conservation of the natural world around them and its biodiversity and how has that changed with the current socioeconomic changes?" The findings indicate that if even one tree species served a useful purpose, then villagers would seek it out and use it. However, the same species may be considered a weed in agricultural land. That is, a particular individual may be involved in conflicting actions regarding a particular species (helpful, harmful, and neutral). Within a village, the knowledge of villagers related to natural resources varied with some having more knowledge and others having less. Some villagers were engaged in horticulture, while others had shifted to placing more emphasis on activities that generated a cash income. This is a type of internal community diversity, in which a "traditional community" is not as homogeneous as some researchers might have assumed in the past. Additionally, this viewpoint is important if conservationists and personnel of related conservation agencies who live and work outside the community want to integrate local people and their traditional environmental knowledge into planned conservation practices. The findings of this study generally suggest that the villagers unintentionally used their subsistence lifeways and land uses to conserve or create biodiversity. However, the descriptive rules used in the community, such as managing white beech or sago palm, seem to somehow be a form of voluntary conservation within the community. Note that the attitude of these people toward conservation or biodiversity is also not homogeneous but is quite diverse within the culture, communities, or even among individuals.

10.5 Internal Diversity and External Inputs

The findings and discussion so far have clarified key answers to general questions such as "What kinds of factors have allowed these communities to maintain lifeways that support ecological sustainability versus acting in a way that might cause a collapse of biocultural diversity in the Solomon Islands?" Successful strategies for harmonizing community welfare and biodiversity will probably include (1) the use of diversified strategies to allow villagers to continue to receive benefits from ecosystem services, (2) the strategy that balances both low-risk and risk-prone methods of earning income, (3) conservation of biodiversity and land in a method that supports these strategies, (4) building and continuing consensus related to the sustainable management and wise use of the forest, and (5) integrating traditional leadership and knowledge for achieving these conditions. However, more discussion is necessary to answer the last question—"How can the societies of Solomon Islands achieve harmonious lifeways that balance the need for conservation and human needs under changing socioeconomic conditions?"—because changes occur rapidly and the people in the urban village are already experiencing growing pains from the expanding population and changing society.

First, we need to acknowledge and respect the internal diversity. As one aspect, internal diversity refers to differences in villagers such as genealogical lineage (relevant to social class and succession of traditional knowledge), educational background (traditional or through Western-style school systems), experience, skills, and attitudes toward development (Chap. 7). Internal diversity can also be observed in each individual because on some occasions an individual can put a value on a specific plant species while at the same time the same individual recognizes the same species as harmful and destroys it in other occasions (Chap. 6). Overall, internal diversity is manifested as a rich community structure and decision-making system that has elements of human ecology. Note that this is the total opposite of the concept to individualization. As shown in the diversity of ethnobotanical knowledge among individual villagers (Chap. 7), the knowledge and competence varied according to social class, i.e., internal diversity in the rural village, because some people consulted a very competent villager every time they needed to know some specific information and the knowledge was a free access "communal property," if not "shared," in the community. In the urban village, individuals had similar levels of competence with little person-to-person variation; this was interpreted as the result of individualization and protection of knowledge as intellectual property and of the influence of official education. Diversity is not cultivated in the individualistic community.

Second, from the viewpoints of sustainability of human activities and environmental conservation, inter-household differences in resource use can be reduced based on the idea of environmental justice, which is recognized to imply equal access to the local resources or environmental benefits (Ohtsuka 2004; Schlosberg and Schlosberg 1999).

Third, external support is needed to empower the community with internal diversity toward harmonized interaction between humans and the natural world (Hviding 2003b). The recommendations of an outside agent might have distinct advantages. Previous success related to the protection of marine resources implies that the integration of indigenous knowledge and customary governance is needed if people are to be expected to follow a protection program (Aswani and Hamilton 2004; Aswani et al. 2007). This study found that people recognized the economic (use for building and tools) and ecological importance (helping birds) of large tree species, such as *Vitex* and *Calophyllum*, which were also targets of commercial logging, although they were likely to weed and hurt young trees of these species that grew in their gardens (Chap. 7). However, informants insisted that they had started to protect these species, following the recommendations of a forestry officer, and they then recognized their potential loss if these trees were to be eliminated and the importance of using them sustainably. In this case, an outside policy maker should try to find methods to encourage everyone to accept all the recommendations or regulations. People's knowledge or recognition apparently appeared to be "environmentally sound," but simultaneously contained some destructive aspects. A method should be carefully developed that considers both the needs of the people and the need for ecologically based conservation.

Fourth, the inhabitants need to be educated on how to cope with the poor conditions created by the overexploitation of resources. However, note that Dunde inhabitants recently participated in and hosted an environmental conservation program (Basi 2008). This may be a first step in allowing the people to find a harmonized method of ensuring both rural development and community welfare.

Fifth, responses to external negative impacts on the island's resources may also cause the villagers to think about how they can manage the land sustainably. The local people have already shown a desire to participate in resource exploitation and have already considered the importance of resources from a Western perspective. According to informants, mangrove was abundant in Koqu Kalena where the people collect crabs and shells. When a commercial logging operation started, the villagers found that the mangrove was so polluted with eroded soils that they demanded the company to stop operations near that area. Recognition of the importance of timber trees, as discussed above, was also a result of confronting resource exploitation. Ecological links between plants and animals were also revealed when the people found those links had been broken, such as in the story of the informant in the Preface. These external factors are thought to play protective roles against the adverse impact of socioeconomic changes, and this is in harmony with the concept of internal diversity. That is, external effects were also diverse and included both direct and indirect effects.

Today, global debates and initiatives tend to agree with this type of proposition. The Convention on Biological Diversity is addressing the rights of indigenous people to use their ecosystems to improve their lifestyles while achieving the goal of biodiversity conservation (Adams et al. 2004; Myers et al. 2000; Sanderson and Redford 2004; van Oudenhoven et al. 2010). Moreover, the Strategic Plan for Biodiversity 2011–2020 (e.g., Target 3) and the SATOYAMA Initiative in the Convention on Biodiversity-10th Convention of Parties, 2010, concluded that biodiversity conservation should aim to preserve sustainable human-modified natural environments, by encouraging broader global recognition of their value. The need to provide people with cash incentives is unavoidable; even though people are enjoying ecosystem services, they need cash today. The cash is helpful in that it allows people to afford an increasing population in a limited land area (Ohtsuka et al. 1995). Accordingly, other recent global forest conservation initiatives based on cash incentives, such as REDD+ (reducing emissions from deforestation and forest degradation in developing countries and the role of conservation, sustainable management of forests and the enhancement of forest carbon stocks in developing countries) or payment for ecosystem services, may play important roles, as long as they are conducted in accordance with the internal diversity. The Solomon Islands displayed very high forest cover until today, even though they had experienced various difficulties in environmental conservation. Lessons from the Solomon Islands will be useful for other societies.

Epilogue

While conducting my fieldwork from 2001 to 2011, I was confused because, on one occasion, some villagers had insisted that logging had adverse effects on ecosystems (e.g., the effects of soil erosion on corals), but a few years later, the same villagers were working for the logging company that caused that damage. I also felt that their knowledge and reasoning related to their experience with the adverse effects of logging, as they had described them to me, sounded thin and superficial, as if they simply repeated what they had heard from external sources (such as government officers or conservation practitioners). I felt that their explanation of the adverse effects was less substantial than their knowledge relating to the ecological connections between plants and animals.

Through this book, I believe I have revealed how the knowledge and behavior of Roviana's people contributed to the mismanagement and degradation of the ecosystem, as well as how the conflicting ideas the people face (survival versus environmental damage) and the effects of this land management somewhat unexpectedly resulted in the development of the area's rich biocultural diversity. I have also clarified how this interaction of humans with the natural world is now at risk: the current pace of socioeconomic change is new to these people and their world is undergoing rapid changes. They have not adapted to these changes by accumulating related knowledge and experience or by developing long-term strategies of adaption. This leaves them at a disadvantage. Nevertheless, the currently ongoing conditions contributing to this change, such as global change or changes in local attitudes, are affecting the local societies and local ecosystems more than in the past. I expect that rural people will develop a strategy allowing them to adapt to change, and I believe the impacts of urban people are already mitigating current adverse effects.

Finally, I would like to introduce the opinion of a young male participant who once worked for the logging company and enjoyed the benefits of having cash income from his work.

> "One day, a foreign tourist visiting our territory happened to meet me. After we talked about my life here in Roviana, he asked me 'Why do you accept the logging of your homeland?' He also strongly addressed the issue, stating 'Forests are so important that you

T. Furusawa, *Living with Biodiversity in an Island Ecosystem*, Ethnobiology,
DOI 10.1007/978-981-287-904-2

> should stop working as a logger immediately.' I could not answer him because I failed to recognize the importance of the forest and to understand his point-of-view at that time. Since then, however, after spending a lot of time thinking about what he said, I gradually came to acknowledge the real importance of the forest for us."

His awakening to the importance of the forest symbolizes the Roviana people's lack of awareness of the impacts of forest exploitation. As researchers, we also desire to assist them in this awakening.

Appendix

Appendix 1: List of plants used in the Roviana, sorted alphabetically by the Roviana name, with observed uses and purposes

Roviana name	Scientific name	Family	Plant type [a]	Use/ purpose [b]	Forest class [c]
Agana	*Pandanus* spp.	Pandanaceae	pl/tr	T	MM
Agana pinomo	*Pandanus* sp.	Pandanaceae	pl/tr	T	
Aroso	*Calamus* spp.	Arecaceae	cl/pl	B, T	
Aroso inoko	*Calamus* sp.	Arecaceae	cl/pl	B	
Asama	*Lygodium* spp.	Schizaeaceae	fn/cl	T	
Babageva masa	*Heritiera littoralis* Ait.	Sterculiaceae	tr	T	MR, MM
Balusa	*Ochroma pyramidale* Urb.	Bombacaceae	tr-l	T	
Bebea	*Tournefortia argentea* L.	Boraginaceae	sh	T	
Bekoto	Various small palms	Arecaceae	pl	B	
Beti	*Bambusa* spp.	Poaceae	gr/ tr-s	B, T	
Binisi	*Phaseolus vulgaris* L.	Fabaceae	hb	Fd	
Binisi noki	*Trichosanthes cucumerina* L.	Cucurbitaceae	hb/cl	Fd	
Bobogele	*Pemphis acidula* J. R. & G. Forst.	Lythrales	sh/ tr-s	B, T	
Bobopa	*Epipremnum* sp.	Araceae	cl	M	

(continued)

T. Furusawa, *Living with Biodiversity in an Island Ecosystem*, Ethnobiology,
DOI 10.1007/978-981-287-904-2

Roviana name	Scientific name	Family	Plant type [a]	Use/ purpose [b]	Forest class [c]
Bolava	*Haplolobus canarioides* Leenh.	Burseraceae	tr	B, Fl	MP, MR, ML
Boroboro	*Coleus* spp.	Lamiaceae	hb	(T)	
Bosi	*Euodia salomonensis* Merr. & Perry	Rutaceae	tr	B	MR, MS, BS
Bosi suka	*Euodia elleryana* Muell.	Rutaceae	tr	B, Fl	MR, MS
Bou	*Fagraea gracilipes* A. Gray	Rubiaceae	tr	B, T	
Buni	*Calophyllum* spp.	Clusiaceae	tr	B, T, Fl	MP, MR, ML
Dadao	*Barringtonia asiatica* (L.) Kurz	Lecythidaceae	tr	T	
Dalou	*Pandanus* sp.	Pandanaceae	pl/tr	T	
Deri	*Citrullus lanatus* (Thunb.) Mansf.	Cucurbitaceae	hb/cr	Fd	
Dikidiki	*Dioscorea esculenta* (Lour.) Burk.	Dioscoreaceae	hb/cl	Fd	
Dodoru	*Trema orientalis* (L.) Bl.	Ulmaceae	tr	B	MS
Domu	Unidentified		tr	B, Fl	
Edeve	*Metroxylon* spp.	Arecaceae	pl	(Fd), B, T	
Eehara	*Horsfieldia spicata* (Roxb.) Sinclair	Myristicaceae	tr	B	MP, MR, ML, BP, BS
Egipalanti	*Solanum melongena* L.	Solanaceae	hb/sh	Fd	
Elelo bakua	*Cassia alata* L.	Caesalpiniaceae	sh	M	
Elohilu	Unidentified		tr	B	
Gaekubo	*Garcinia celebica* L., *G. solomonensis* A. C. Sm.	Clusiaceae	tr	B	MP, MR, MS, ML
Geholo	*Schleinitzia novo-guineensis* (Warb.) Verdc.	Leguminoceae	tr-s	B, Fl	
Gozigolo	*Scindapsus altissimus* V. A. V. R.	Araceae	cl	T	
Gurata	*Morinda citrifolia* L.	Rubiaceae	tr-s	M, Fl	

(continued)

Roviana name	Scientific name	Family	Plant type [a]	Use/ purpose [b]	Forest class [c]
Habe	Unidentified			T	
Haila	*Syzygium* sp.	Myrtaceae	tr	T	BP
Hakua	*Musa* spp.	Musaceae	hb/ tr-s	Fd	
Halagire	Pandanaceae			T	BS
Hame	*Calophyllum paludosum* C. T. White., *C. neo-ebudicum* Guill.	Clusiaceae	tr	B, Fl	MR
Harekete	*Microsorium scolopendria* (Burm. f.) Copel.	Polypodiaceae	fn/cl	M	
Hebere	*Dillenia ingens* Burtt	Dilleniaceae	tr-l	B, Fl	MR, MS, ML
Heta (heta manavasa)	*Areca catechu* L.	Arecaceae	pl	Fd, M, B	
Heta pinomo	*Areca macrocalyx* Zipp. Ex Bl.	Arecaceae	pl	(Fd), B, Fl	
Hioko	*Syzygium* sp.	Myrtaceae	tr	B	MP, MR, MS, BP, BS
Hipahipala	*Syzygium* sp.	Myrtaceae	tr	B	
Hivu	*Acalypha grandis* Benth.	Euphorbiaceae	tr/sh	Fl	
Horehore	*Calophyllum* sp.	Clusiaceae	tr/tr-l	B	MP, MR, MS, ML
Hovaka	*Burckella obovata* (Forst.) Pierre	Sapotaceae	tr-l	B, T	
Huhu	*Ficus septica* Burm. f.	Moraceae	tr	Fl	MLBS
Igisi	*Piper betle* L.	Piperaceae	sh/cl	Fd, M	
Iuki	*Eucalyptus deglupta* Blume	Myrtaceae	tr	Fl	
Ivili	*Intsia bijuga* (Colebr.) Kuntze	Caesalpiniaceae	tr-m	B, T	
Kabisi	*Brassica chinensis* L.	Cruciferae	hb	Fd	
Kakarumu	*Lumnitzera littorea* (Jack.) Voigt	Combretaceae	tr-s	B, T	

(continued)

Roviana name	Scientific name	Family	Plant type [a]	Use/ purpose [b]	Forest class [c]
Kaŋana	*Gulubia macrospadix* (Burret) H. E. Moore	Arecaceae	pl	B	MP, MR, MS, ML
Kapuhu	*Dillenia salomonensis* (C. T. White) Hoogl.	Dilleniaceae	tr	B, T, Fl	MRML
Karuvera	*Alocasia macrorrhiza* (L.) G. Don	Araceae	hb		
Keto	*Zea mays* L.	Poaceae	gr/hb	Fd	
Kikilapa (kilala)	*Ochroma pyramidale* Urb.	Bombaceae	tr-l	T	
Kikopo	*Buchanania* sp.	Anacardiaceae	tr/tr-l	Fl	
Kinu	*Barringtonia procera* (Miers) R.Knuth	Lecythidaceae	tr	Fd	
Kokeqolo	*Aglaia brassii* Merr. and Perry	Meliaceae	tr	B	MP, MR, ML, BP, BS
Kokolomana	*Sterculia shillinglawii* Muell.	Sterculiaceae	tr	Fl	BS
Kosikosiri	*Diplazium esculentum* (Retz.) Sw.	Athyriaceae	fn	Fd	
Kuava	*Psidium guajava* L.	Myrtaceae	tr-s	Fd, T	
Kukaba	*Cucumis sativus* L.	Cucurbitaceae	hb/cl	Fd	
Kureu	*Diospyros ferrea* (Willd.) Bakh.	Ebenaceae	tr-s	B	MP, MR
Laini	*Citrus aurantifolia* (Christm.) Swing.	Rutaceae	tr-s	Fd, M	
Lemone	*Citrus limon* (L.) Burm. f.	Rutaceae	tr-s	(Fd), M	
Leqe	*Gnetum gnemon* L.	Gnetaceae	tr	Fd	BS
Levaleva	Unidentified		tr	B	ML
Likisi	*Allium porrum* L.	Liliaceae	hb	Fd	
Liqeliqe	Unidentified		tr	B	ML, BP
Lozi	*Ceiba pentandra* (L.) Gaertn., *Bombax malabaricum* DC.	Bombaceae	tr-m	T	

(continued)

Roviana name	Scientific name	Family	Plant type [a]	Use/ purpose [b]	Forest class [c]
Lulua	*Amoora cucullata* Roxb.	Meliaceae	tr	B	MP, MR, BP, BS
Luluzu	*Mikania cordata* (Burm. f.) B. L. Rob.	Asteraceae	hb/cl	M	
Luzu vaka	*Ipomoea batatas* (L.) Lam.	Convolvulaceae	hb/cr	Fd, M	
Mahigeli	*Gulubia hombronii* Becc.	Arecaceae	pl	B	
Manioko	*Carica papaya* L.	Caricaceae	tr	Fd	
Marihi	*Dioscorea* spp.	Dioscoreaceae	hb/cl	Fd	
Mavuana	*Flueggea flexuosa* Müll. Arg.	Euphorbiaceae	tr	B, T, Fl	BS
Mokulou	Unidentified		tr	B	
Naqarita	*Cananga odorata* (Lamk.) Hook. f. & Thoms.	Annonaceae	tr-m	M, T	
Naru	*Casuarina equisetifolia* J. R. & G. Forst.	Casuarinaceae	tr-m	T, Fl	
Natoŋo	*Rhus taitensis* Guill.	Anacardiaceae	tr	B, T, Fl	MS
Neka	*Hibiscus manihot* L.	Malvaceae	sh	Fd	
Nekete	*Pipturus argenteus* (Forst. f.) Wedd.	Urticuliaceae	sh/ tr-s	M	
Nobinobi ime	Unidentified	Fabaceae		T, Fl	
Ŋohara	*Cocos nucifera* L.	Arecaceae	pl/ tr-m	Fd, M, B, T, (Fl)	MM
Nonoqara	*Nephrolepis hirsutula* (Forst.) Presl	Oleandraceae	fn	T	
Okete	*Canarium indicum* L.	Burseraceae	tr-m	Fd	
Okokete	*Garuga floribunda* Dence., *Canarium vitiense* A. Grat	Burseraceae	tr/ tr-m	B	MR, MS
One	*Weinmannia blumei* Planch.	Cunoniaceae	tr	Fl	MP, MR, ML
Onioni	*Allium cepa* L. var. *aggregatum* G. Don	Amaryllidaceae	hb	Fd	

(continued)

Roviana name	Scientific name	Family	Plant type [a]	Use/ purpose [b]	Forest class [c]
Opiti	*Spondias dulcis* Sol. ex Parlk., *Averrhoa carambola* L.			Fd	
Ore marihi	*Manihot esculenta* Crantz	Euphorbiaceae	sh	Fd	
Paenapo	*Ananas comosus* (L.) Merr.	Bromeliaceae	hb	Fd	
Pakopako	*Phyllanthus ciccoides* Muell. Arg.	Euphorbiaceae	tr	Fl	MS
Paloto	*Palaquium erythrospermum* Lam.	Sapotaceae	tr	B, Fl	MR, ML
Pamuken	*Cucurbita moschata* Duchesne	Cucurbitaceae	hb/cl	Fd	
Pate	*Pandanus tectorius* Park.	Pandanaceae	pl/tr	T	
Pepa	*Capsicum annum* L. var. *grossum* Sendt.	Solanaceae	hb	Fd	
Pepeo	*Terminalia brassii* Exell	Combretaceae	tr-l	B	
Petepete	*Litsea domarensis* Schmidt	Lauraceae	tr	B	MS, ML
Petu	*Bruguiera gymnorhiza* (L.) Lam.	Rhizophoraceae	tr	Fd, B, T, Fl	MM
Petukele	*Macaranga fimbriata* S. Moore	Euphorbiaceae	tr	B, Fl	
Pidiki	*Syzygium onesimum* Merr. & Perry	Myrtaceae	tr	B, T, Fl	MP, MR, ML, BP
Pike	*Horsfieldia polyantha* Warb.	Myristicaceae	tr	Fl	ML, BP, BS
Pilasi	Unidentified			T	
Pinati	*Arachis hypogaea* L.	Fabaceae	hb	Fd	
Pokopoko	*Campnosperma brevipetiolata* Volkens	Anacardiaceae	tr-l	B, T, Fl	MR, ML
Pomolo	*Citrus grandis* (L.) Osbeck	Rubiaceae	tr-s	Fd	

(continued)

Roviana name	Scientific name	Family	Plant type [a]	Use/ purpose [b]	Forest class [c]
Qema	*Pometia pinnata* Forst. f.	Sapindaceae	tr-m	B, T, Fl	MR, MS, BP
Qoliti	*Gmelina moluccana* Backer ex K.Heyne	Verbenaceae	tr	T	MS
Rapa	*Ficus lancibracteata* Corner	Moraceae	tr	B, Fl	MR, BS
Rereveti	*Guettarda speciosa* L.	Rubiaceae	tr	Fl	
Riqi	*Pterocarpus indicus* Willd.	Fabaceae	tr-m/ tr-l	B, T	
Ruqupole	*Ocimum basilicum* L.	Lamiaceae	hb/ ssh	(Fd), T	
Saladia	*Lactuca sativa* L.	Asteraceae	hb	Fd	
Sasopo	*Annona muricata* L.	Annonaceae	tr-s	Fd	
Sili	*Capsicum frutescens* L.	Solanaceae	hb/ ssh	Fd	
Sosoruku	*Sterculia shillinglawii* Muell.	Sterculiaceae	tr	T	
Suri	*Diospyros* sp., *Timonius forsteri* DC		sh/ tr-s	B, T, Fl	MP, MR, ML, BS
Suti	*Saccharum officinarium* L.	Poaceae	gr/hb	Fd	
Talo	*Colocasia esculenta* (L.) Schott.	Araceae	hb	Fd	
Tamata	*Lycopersicon lycopersicum* (L.) Karst.	Solanaceae	hb	Fd	
Tatalise	*Terminalia catappa* L.	Combretaceae	tr-m	M	
Tila	*Haplolobus floribundus* (Schum.) Lam.	Burseraceae	tr-m	Fl	MP, MR, ML
Tita	*Parinari glaberrima* (Hassk.) Hassk.	Chrysobalanaceae	tr	B, Fl	MP, MR, ML
Titimunuhaha	*Clerodendrum buchananii* (Roxb.) Walp.	Verbenaceae	sh/ tr-s	M	
Tivativa	*Syzygium* sp.	Myrtaceae	tr	T	MP

(continued)

Roviana name	Scientific name	Family	Plant type [a]	Use/ purpose [b]	Forest class [c]
Tobo	*Macaranga lanceolata* Pax. & Hoffm.	Euphorbiaceae	tr	Fl	MR, MS, ML, BS
Toqoŋeta	*Timonius timon* (Spreng.) Merr.	Rubiaceae	tr	M, B, T, Fl	
Tototu	*Sonneratia caseolaris* (L.) Engl.	Sonneratiaceae	tr	B, T	MM
Totuana	*Alstonia spectabilis* R. Br.	Apocynaceae	tr	M, B, Fl	MS, BP
Tovinia	*Canarium salomonense* Burtt	Burseraceae	tr-m	Fd, Fl	
Tukituki	*Macaranga* spp.	Euphorbiaceae	tr	B, T, Fl	MS, ML, BS
Turutoŋoro	*Horsfieldia irya* (Gaertn.) Warb.	Myrtaceae	tr	B	MR, ML, BS
Uotakuresu	*Nasturtium officinale* R. Br.	Brassicaceae	hb	Fd	
Valo	*Gonystylus macrophyllus* (Miq.) A. Shaw, *G. megacarpus* C. T. White	Thymelaeaceae	tr	B	MP, MR, MS, ML
Varu	*Hibiscus tiliaceus* L.	Malvaceae	tr	T, Fl	BS
Vasara	*Vitex cofassus* Reinw. ex Bl.	Verbenaceae	tr-l	B, T, Fl	BP, BS
Vasavasara	*Geniostoma rupestris* J. R. & G. Forst.	Loganiaceae	tr	B	MS, BP
Vogi	*Crinum asiaticum* L.	Amaryllidaceae	hb	M	
Voko	*Ptychosperma salomonense* Burret	Arecaceae	pl	T	
Vorusu	*Ceriops tagal* (Pers.) C. B. Rob.	Rhizophoraceae	tr	B	
Vosevose	*Neonauclea* spp.	Naucleaceae	tr	B, T, Fl	BP, BS,
Vuagore	*Dysoxylum excelsum* Bl.	Meliaceae	tr	T	BS

(continued)

Roviana name	Scientific name	Family	Plant type [a]	Use/ purpose [b]	Forest class [c]
Zamara	*Commersonia bartramia* (L.) Merr.	Sterculiaceae	tr-s	B, Fl	MS
Zazalagaba	*Codiaeum variegatum* Blume	Euphorbiaceae	sh	(T)	
Zipolo	*Cordyline terminalis* Kunth	Agavaceae	sh	(T)	
Zizito	*Galearia celebica* Koord.	Euphorbiaceae	tr-m	(M), Fl	
Zovi	*Premna corymbosa* (Burm. f.) R. & W.	Verbenaceae	tr-s	T	
No name (introduced plant)	*Catharanthus roseus* (L.) G. Don	Apocynaceae	hb	M	
Name unidentified 1				B	
Name unidentified 2				B	
Name unidentified 3				B	
Name unidentified 4				B	
Name unidentified 5				B	
Name unidentified 6				B	
Name unidentified 7				B	
Name unidentified 8	Unidentified			B	

[a]Plant type: *cl* climber, *cr* creeper, *ep* epiphyte, *fn* fern, *gr* grass, *hb* herb, *sh* shrub, *tr* tree (size unidentified), *tr-s* tree-small (<12 m tall), *tr-m* tree-medium (12–25 m tall), *tr-l* tree-large (>25 m tall) (Henderson and Hancock 1988)

[b]Use purpose: *Fd* food, *M* medicine, *B* building, *T* tool (including magical purposes), *Fl* fuel; uses/ purposes observed during the research periods are shown; those observed outside of the study period were also shown in parenthesis

[c]Forest class: *MP* main island, primary forest, *MR* main island, reserve forest, *MS* main island, secondary forest, *ML* main island, logged forest, *BP* barrier island, primary forest, *BS* barrier island, secondary forest, *MM* main island, mangrove; only trees observed in this study were shown

References

Adams WM, Aveling R, Brockington D, Dickson B, Elliott J, Hutton J, Roe D, Vira B, Wolmer W (2004) Biodiversity conservation and the eradication of poverty. Science 306(5699):1146–1149

Akimichi T (1978) The ecological aspect of Lau (the Solomon Islands) ethnoichthyology. J Polyn Soc 87(4):301–326

Akimichi T (1991) Sea tenure and its transformation in the Lau of north Malaita, Solomon Island. S Pac Study 12(1):7–22

Allaby R (2007) Origins of plant exploitation in near Oceania: a review. In: Friedlaender JS (ed) Population genetics, linguistics and culture history in the South West Pacific. Oxford University Press, New York, pp 181–98

Allen M, Sinclair D (2013) The Solomon Islands in transition. In: State, society and governance in Melanesia in brief 2013/5. The Australian National University, Canberra

Asian Development Bank (2007a) Key indicators 2007: inequality in Asia. Asian Development Bank, Manila

Asian Development Bank (2007b) ADB Social and Poverty Assessment: Solomon Islands Emergency Assistance Project. Asian Development Bank, Manila

Aswani S (1997) Customary sea tenure and artisanal fishing in the Roviana and Vonavona Lagoons, Solomon Islands. The evolutionary ecology of marine resource utilization. Ph. D. dissertation, University of Hawaii at Manoa

Aswani S (1999) Common property models of sea tenure: a case study from the Roviana and Vonavona Lagoons, New Georgia, Solomon Islands. Hum Ecol 27(3):417–453

Aswani S (2000) Changing identities: the ethnohistory of Roviana predatory head-hunting. J Polyn Soc 109(1):39–70

Aswani S (2002) Assessing the effects of changing demographic and consumption patterns on sea tenure regimes in the Roviana Lagoon, Solomon Islands. Ambio 31(4):272–284

Aswani S, Furusawa T (2007) Do marine protected areas affect human nutrition and health? A comparison between villages in Roviana, Solomon Islands. Coast Manag 35(5):545–565

Aswani S, Hamilton RJ (2004) Integrating indigenous ecological knowledge and customary sea tenure with marine and social science for conservation of bumphead parrotfish (Bolbometopon muricatum) in the Roviana Lagoon, Solomon Islands. Environ Conserv 31(01):69–83

Aswani S, Lauer M (2006) Incorporating fishermen's local knowledge and behavior into geographical information systems (GIS) for designing marine protected areas in Oceania. Hum Organ 65(1):81–102

Aswani S, Sabetian A (2009) Implications of urbanization for artisanal parrotfish fisheries in the Western Solomon Islands. Conserv Biol 24(2):520–30

T. Furusawa, *Living with Biodiversity in an Island Ecosystem*, Ethnobiology,
DOI 10.1007/978-981-287-904-2

Aswani S, Sheppard P (2003) The archaeology and ethnohistory of exchange in precolonial and colonial Roviana: gifts, commodities, and inalienable possessions. Curr Anthropol 44(S5):S51–S78

Aswani S, Vaccaro I (2008) Lagoon ecology and social strategies: habitat diversity and ethnobiology. Hum Ecol 36(3):325–341

Aswani S, Albert S, Sabetian A, Furusawa T (2007) Customary management as precautionary and adaptive principles for protecting coral reefs in Oceania. Coral Reefs 26(4):1009–1021

Atran S, Medin D, Ross N, Lynch E, Coley J, Ek' EU, Vapnarsky V (1999) Folkecology and commons management in the Maya Lowlands. Proc Natl Acad Sci U S A 96(13):7598–7603

Atran S, Medin D, Ross N, Lynch E, Vapnarsky V, Ek' EU, Coley J, Timura C, Baran M (2002) Folkecology, cultural epidemiology, and the spirit of the commons: a garden experiment in the Maya Lowlands, 1991–2001. Curr Anthropol 43(3):421–450

Balick MJ, Cox PA (1996) Plants, people, and culture: the science of ethnobotany. Scientific American Library, New York

Barclay K (2004) Political economy of identities in an instance of globalisation: history of a Solomon Islands Japanese Joint Venture Tuna Fishing Corporation (1971–2000). Working paper (Australian National University. Resource Management in Asia-Pacific). Australian National University, Canberra

Barclay K, Wakabayashi Y (2000) Solomon Taiyo Ltd: Tuna dreams realized? Pac Econ Bull 15(1):34–47

Basi J (2008) Munda celebrates "Conservation Day". Solomon Times Online. 11 August

Bayliss–Smith T (1974) Constraints on population growth: the case of the Polynesian Outlier Atolls in the precontact period. Hum Ecol 2(4):259–295

Bayliss–Smith T (2006) Fertility and the depopulation of Melanesia: childlessness, abortion and introduced disease in Simbo and Ontong Java, Solomon Islands. In: Ulijaszek SJ (ed) Population, reproduction and fertility in Melanesia. Berghahan Books, New York, pp 13–51

Bayliss-Smith T, Hviding E (2012) Irrigated taro, malaria and the expansion of chiefdoms: ruta in New Georgia, Solomon Islands. In: Spriggs M, Addison D, Matthews P (eds) Irrigated cultivation of Colocasia esculenta in the Indo-Pacific: biological, social and historical perspectives, Senri ethnological studies, 78. National Museum of Ethnology, Osaka, pp 219–254

Bayliss–Smith T, Hviding E, Whitmore T (2003) Rainforest composition and histories of human disturbance in Solomon Islands. Ambio 32(5):346–352

Bélair C, Ichikawa K, Wong BYL, Mulongoy KJ (eds) (2010) Sustainable use of biological diversity in socio-ecological production landscapes: background to the "Satoyama initiative for the benefit of biodiversity and human well–being". Secretariat of the Convention on Biological Diversity, Montreal

Bennett JA (1987) Wealth of the Solomons: a history of a Pacific Archipelago, 1800–1978. University of Hawaii Press, Honolulu

Bennett JA (2000) Pacific forest: a history of resource control and contest in Solomon Islands, c.1800 – 1997. White Horse Publisher, Cambridge

Bennett BC, Prance GT (2000) Introduced plants in the indigenous Pharmacopoeia of Northern South America. Econ Bot 54(1):90–102

Benz BF et al (2000) Losing knowledge about plant use in the Sierra de Manantlan biosphere reserve, Mexico. Econ Bot 54(2):183–191

Blaber SJM (1991) A checklist and notes on the current status of the birds of New Georgia, Western Province, Solomon Islands. Emu 90(4):205–214

Borgatti SP (1996) ANTHROPAC 4.0. Analytic Technologies, Natick

Boster JS (1986) Exchange of varieties and information between Aguaruna Manioc cultivators. Am Anthropol 88(2):428–436

Braithwaite J, Dinnen S, Allen M, Braithwaite V, Charlesworth H (2010) Pillars and shadows: statebuilding as peacebuilding in Solomon Islands. ANU E Press, Canberra

Brooks TM, Mittermeier RA, da Fonseca GAB, Gerlach J, Hoffmann M, Lamoreux JF, Mittermeier CG, Pilgrim JD, Rodrigues ASL (2006) Global biodiversity conservation priorities. Science 313(5783):58–61

Bulmer R (1982) Traditional conservation practices in Papua New Guinea. In: Morauta L, Pernetta J, Heaney W (eds) Traditional conservation in Papua New Guinea: implications for today. Institute of Applied Social and Economic Research, Boroko, pp 59–77
Campbell IC (1989) A history of the Pacific Islands. University of California Press, Berkeley
Caniago I, Siebert Stephen F (1998) Medicinal plant ecology, knowledge and conservation in Kalimantan, Indonesia. Econ Bot 52(3):229–250
Carneiro RL (1964) Shifting cultivation among the Amahuaca of Eastern Peru. In: Becher H (ed) Beitrage zur Volkerkunde, Sudimerikas, Volkerkundliche Abhandlungen, Band 1. Kommisionsverlag Münstermann-Druck, Hanover, pp 9–18
Carrier JG (2010) Protecting the environment the natural way: ethical consumption and commodity fetishism. Antipode 42(3):672–689
Carroll V (1975) Pacific atoll populations. University of Hawaii Press, Honolulu
Carter B (1997) Conservation strategy for the Island of Tetepare. Report prepared by Bill Carter with the Assistance of Friends of Tetepare and WWF South Pacific Program. WWF, Gizo
Case RJ, Pauli GF, Doel Soejarto D (2005) Factors in maintaining indigenous knowledge among ethnic communities of Manus Island. Econ Bot 59(4):356–365
Chivian E, Bernstein A (2008) Sustaining life: how human health depends on biodiversity. Oxford University Press, New York
Cocklin C, Keen M (2000) Urbanization in the Pacific: environmental change, vulnerability and human security. Environ Conserv 27(04):392–403
Coelli T, Fleming E (2004) Diversification economies and specialisation efficiencies in a mixed food and coffee smallholder farming system in Papua New Guinea. Agric Econ 31(2–3):229–239
Cohen P, Tapala S, Rikio A, Kukiti E, Sori F, Hilly Z, Alexander TJ, Foale S (2014) Developing a common understanding of taxonomy for fisheries management in North Vella Lavella, Solomon Islands. SPC Tradit Mar Resour Manag Knowl Inf Bull 33:3–12
Conklin HC (1972) Hanunoo agriculture in the Philippines, Food and Agriculture Organization (FAO) forestry department paper, 12. FAO, Rome
Conservation International (2012) Overview: East Melanesian Islands: the hotspots Asia-Pacific, Conservation International. http://www.conservation.org/where/priority_areas/hotspots/asia-pacific/East-Melanesian-Islands/Pages/default.aspx. Accessed 5 Feb 2014
Dennett G, Connell J (1988) Acculturation and health in the highlands of Papua New Guinea: dissent on diversity, diets, and development. Curr Anthropol 29(2):273–299
Diamond JM (1988) Express train to Polynesia. Nature 336:307–308
Dignan CA, Burlingame B, Kumar S, Aalbersberg W (1994) The Pacific Islands food consumption tables. New Zealand Institute for Crop and Food Research, Palmeston North
Division RD (2001) Western province development profile. Ministry of Provincial Government and Rural Development, Honiara
Doughty C, Day N, Plant A (1999) Birds of the Solomons, Vanuatu & New Caledonia. A & C Black, London
Drewnowski A, Popkin BM (1997) The nutrition transition: new trends in the global diet. Nutr Rev 55(2):31–43
Dwyer PD (1990) The pigs that ate the garden: a human ecology from Papua New Guinea. University of Michigan Press, Ann Arbor
Dwyer PD (1994) Modern conservation and indigenous peoples: in search of wisdom. Pac Conserv Biol 1:91–97
Eason RJ, Pada J, Wallace R, Henry A, Thornton R (1987) Changing patterns of hypertension, diabetes, obesity and diet among Melanesians and Micronesians in the Solomon Islands. Med J Aust 146(9):465–469
Ellen R (1982) Environment, subsistence and system: the ecology of small-scale social formations. Cambridge University Press, Cambridge
Fazey I, Pettorelli N, Kenter J, Wagatora D, Schuett D (2011) Maladaptive trajectories of change in Makira, Solomon Islands. Glob Environ Chang 21(4):1275–1289

Feeny D, Berkes F, McCay BJ, Acheson JM (1990) The tragedy of the commons: twenty-two years later. Hum Ecol 18(1):1–19

Filer C (2004) A serious case of conditionality: The World Bank gets stuck in the forests of PNG. Dev Bulletin (Canberra) 65:95–100

Filer C (2011) Interdisciplinary perspectives on historical ecology and environmental policy in Papua New Guinea. Environ Conserv 38:256–269

Foale S (2001) 'Where's our development?' Landowner aspirations and environmentalist agendas in Western Solomon Islands. Asia Pac J Anthropol 2(2):44–67

Foale S, Manele B (2004) Social and political barriers to the use of marine protected areas for conservation and fishery management in Melanesia. Asia Pac Viewp 45(3):373–386

Foale S, Cohen P, Januchowski-Hartley S, Wenger A, Macintyre M (2011) Tenure and taboos: origins and implications for fisheries in the Pacific. Fish Fish 12(4):357–369

Food and Agriculture Organization (2011) State of the world's forests 2011. Food and Agricultural Organization of the United Nations, Rome

Food and Agriculture Organization/World Health Organization/United Nations University (2004) Human energy requirements: report of a Joint FAO/WHO/United Nations University Expert Consultation, FAO food and nutrition technical report series 1. Food and Agriculture Organization, Rome

Fratkin EM (1998) Ariaal pastoralists of Kenya: surviving drought and development in Africa's arid lands. Allyn & Bacon, Boston

Frazer I (1997) The struggle for control of Solomon Island forests. Contemp Pac 9(1):39–72

Friedlaender JS, Friedlaender FR, Reed FA, Kidd KK, Kidd JR, Chambers GK, Lea RA, Loo J-H, Koki G, Hodgson JA, Andrew Merriwether D, Weber JL (2008) The genetic structure of Pacific islanders. PLoS Genet 4(1):e19

Furusawa T (2004) Minzoku Chishiki ni Motoduku Ningen-Shokubutsu-Dobutsu no Kankei [Human-plant-animal relationships analyzed as folk knowledge]. In: Ohtsuka R (ed) Solomon Shoto: Saigo no Nettairin [Solomon Islands: the last frontier of tropical rain forest]. University of Tokyo Press, Tokyo, pp 55–81. (In Japanese)

Furusawa T (2006) The roles of western biomedicine and folk medicine in rural Solomon Islands: a quantitative analysis of villagers' response to illness. Trop Med Health 34(2):83–91

Furusawa T (2009) Changing ethnobotanical knowledge of the Roviana people, Solomon Islands: quantitative approaches to its correlation with modernization. Hum Ecol 37(2):147–159

Furusawa T (2011) Ecological and economic analyses of urbanization in the customary lands of the Solomon Islands. People Cult Ocean 27:1–17

Furusawa T (2012) Tracking fishing activities of the Roviana population in the Solomon Islands using a portable global positioning system unit and a heart rate monitor. Field Methods 24(2):216–229

Furusawa T, Ohtsuka R (2006) Inter-household variations in subsistence strategies within a rural society of Roviana, Solomon Islands: an analysis of agricultural production and cash income in relation to socio–demographic factors. Tropics 15(1):29–37

Furusawa T, Ohtsuka R (2009) The role of barrier Islands in subsistence of the inhabitants of Roviana Lagoon, Solomon Islands. Hum Ecol 37(5):629–642

Furusawa T, Pahari K, Umezaki M, Ohtsuka R (2002) Study on the changing characteristics of land use as commons in Roviana, Western Province, Solomon Islands using high resolution Satellite data and aerial photographs. In: Proceedings of 23rd Asian conference on remote sensing, CD-ROM, Kathmandu, Nepal.

Furusawa T, Pahari K, Umezaki M, Ohtsuka R (2004) Impacts of selective logging on New Georgia Island, Solomon Islands evaluated using very-high-resolution satellite (IKONOS) data. Environ Conserv 31(04):349–355

Furusawa T, Maki N, Suzuki S (2008) Bacterial contamination of drinking water and nutritional quality of diet in the areas of the Western Solomon Islands devastated by the April 2, 2007 earthquake? Tsunami. Trop Med Health 36(2):65–74

Furusawa T, Naka I, Yamauchi T, Natsuhara K, Kimura R, Nakazawa M, Ishida T, Inaoka T, Matsumura Y, Ataka Y, Nishida N, Tsuchiya N, Ohtsuka R, Ohashi J (2010) The Q223R polymorphism in LEPR is associated with obesity in Pacific Islanders. Hum Genet 127(3):287–94

Furusawa T, Furusawa H, Eddie R, Tuni M, Pitakaka F, Aswani S (2011a) Communicable and non–communicable diseases in the Solomon Islands villages during recovery from a massive earthquake in April 2007. N Z Med J 124(1333):17–28

Furusawa T, Naka I, Yamauchi T, Natsuhara K, Kimura R, Nakazawa M, Ishida T, Nishida N, Eddie R, Ohtsuka R, Ohashi J (2011b) The serum leptin level and body mass index in Melanesian and Micronesian Solomon Islanders: focus on genetic factors and urbanization. Am J Hum Biol 23(4):435–44

Furusawa T, Naka I, Yamauchi T, Natsuhara K, Eddie R, Kimura R, Nakazawa M, Ishida T, Inaoka T, Matsumura Y, Ataka Y, Ohtsuka R, Ohashi J (2013) Hypertension-susceptibility gene prevalence in the Pacific Islands and associations with hypertension in Melanesia. J Hum Genet 58:142–9

Furusawa T, Sirikolo MQ, Sasaoka M, Ohtsuka R (2014) Interaction between forest biodiversity and people's use of forest resources in Roviana, Solomon Islands: implications for biocultural conservation under socioeconomic changes. J Ethnobiol Ethnomed 10(1):10

Garonna I, Fazey I, Brown ME, Pettorelli N (2009) Rapid primary productivity changes in one of the last coastal rainforests: the case of Kahua, Solomon Islands. Environ Conserv 36(03):253–260

Goldman M (2003) Partitioned nature, privileged knowledge: community-based conservation in Tanzania. Dev Chang 34(5):833–862

Goodbody S, Humphreys GS (1986) Soil chemical status and the prediction of sweet potato yields. Trop Agric 63:209–211

Gordon RG Jr (2005) Ethnologue: languages of the world, Dallas, Tex.: summer Institute of Linguistics International. Online version: http://www.ethnologue.com. Accessed 5 Feb 2014

Green RE, Cornell SJ, Scharlemann JPW, Balmford A (2005) Farming and the fate of wild nature. Science 307(5709):550–555

Groombridge B, Jenkins MD (2002) World atlas of biodiversity. University of California Press, Berkley/Los Angeles

Hancock IR, Henderson CP (1988) Flora of the Solomon Islands. Dodo Creek Research Station, Honiara

Hardin G (1968) The tragedy of the Commons. Science 162:1243–1248

Hartemink AE (2003) Sweet potato yields and nutrient dynamics after short–term fallows in the humid lowlands of Papua New Guinea. NJAS Wagening J Life Sci 50(3):297–319

Hartemink AE, Bourke RM (2000) Nutrient deficiencies of agricultural crops in Papua New Guinea. Outlook Agric 29(2):97–108

Henderson CP, Hancock IR (1988) A guide to the useful plants of Solomon Islands. Ministry of Agriculture and Lands, Honiara

Hodge AM, Dowse GK, Koki G, Mavo B, Alpers MP, Zimmet PZ (1995) Modernity and obesity in coastal and highland Papua New Guinea. Int J Obes 19(3):154–161

Hviding E (1996) Guardians of Marovo Lagoon: practice, place, and politics in maritime Melanesia. University of Hawaii Press, Honolulu

Hviding E (2003a) Between knowledges: pacific studies and academic disciplines. Contemp Pac 15(1):43–73

Hviding E (2003b) Contested rainforests, NGOs, and projects of desire in Solomon Islands. Int Soc Sci J 55(178):539–554

Hviding E (2005) Kiladi oro Vivineidi ria Tingitonga pa Idere oro pa Goana pa Marovo: reef and rainforest: an environmental encyclopedia of Marovo Lagoon, Solomon Islands. UNESCO, Paris

Hviding E (2006) Knowing and managing biodiversity in the Pacific Islands: challenges of environmentalism in Marovo Lagoon. Int Soc Sci J 58(187):69–85

Hviding E, Bayliss-Smith T (2000) Islands of rainforest: agroforestry, logging and eco–tourism in Solomon Islands. Ashgate, Farnham

Inubushi K, Anzai T (2001) Dojogaku Gairon [An introduction to soil science]. Asakura Shoten, Tokyo. (In Japanese)

Ishimori D (2001) Kastom and fasin: the epistemological relationship between past and present in Vangunu, Solomon Islands. Minzokugaku-kenku Jpn J Ethnol 66(2):222–239

Ishimori D (2007) Disentangling fundamentalism and nativistic movements: an analysis of the Christian Fellowship Church in the Solomon Islands. People Cult Ocean 23:33–52

Jackson KB (1978) Tie Hokara, Tie Vaka: black man, white man: a study of the New Georgia group to 1925. Ph.D. dissertation, Australian National University, Canberra

Kabutaulaka TT (2006) Global capital and local ownership in Solomon Islands' forestry industry. In: Firth S (ed) Globalization and governance in the Pacific Islands. ANU E Press, Canberra, pp 239–257

Kagawa Y (2001) Gotei Shokuhin Seibun Hyo [Standard tables of food composition in Japan, Fifth Edition]. Kagawa Nutrition University, Tokyo. (In Japanese)

Kareiva P, Watts S, McDonald R, Boucher T (2007) Domesticated nature: shaping landscapes and ecosystems for human welfare. Science 316(5833):1866–1869

Kawabe T (1983) Development of hunting and fishing skill among boys of the Gidra in lowland Papua New Guinea. J Hum Ergol 12(1):65–74

Kenter JO, Hyde T, Christie M, Fazey I (2011) The importance of deliberation in valuing ecosystem services in developing countries-evidence from the Solomon Islands. Glob Environ Chang 21(2):505–521

Kimura R, Ohashi J, Matsumura Y, Nakazawa M, Inaoka T, Ohtsuka R, Osawa M, Tokunaga K (2008) Gene flow and natural selection in Oceania human populations inferred from genome-wide SNP typing. Mol Biol Evol 25:1750–1761

King H, Collins AM (1989) A modernity score for individuals in Melanesian society. P N G Med J 32(1):11–22

Kirch PV, Rallu J-L (2007) The growth and collapse of Pacific Island societies: archaeological and demographic perspectives. University of Hawaii Press, Honolulu

Kobayashi S, Toujyou B, Vylayphone A, Gunawan H, Ugarte J, Ishimaru K (2013) Evaluation of carbon credit by semi-domestication of forest resources and safeguard. In: The proceedings of the international workshop on incentive of local community for REDD and semi-domestication of non-timber forest products, pp 179–192

Kwa'ioloa M, Burt B (2001) Na Masu'u kid'i Kwara'ae: Tualaka'i Solomon Islands fa'inia logo na rfi ne'e bulao saena fanoa kia: "Our Forest of Kwara'ae": our life in Solomon Islands and the things which grow in our home. British Museum Press, London

Larmour P (1997) Governance and reform in the South Pacific. Asia Pacific Press, Canberra

Larmour P, Ben W, Zoleveke G (1979) Land in Solomon Islands. Institute of Pacific Studies, University of the South Pacific and the Ministry of Agriculture and Lands, Solomon Islands, Honiara

Lawrence D, Peart DR, Leighton M (1998) The impact of shifting cultivation on a rainforest landscape in West Kalimantan: spatial and temporal dynamics. Landsc Ecol 13:135–48

Lewis WH (2003) Pharmaceutical discoveries based on ethnomedicinal plants: 1985 to 2000 and beyond. Econ Bot 57:126–134

Liloqula R (2000) Understanding the conflict in Solomon Islands as a practical means to peacemaking. Dev Bull 53:40–43

Lum JK, Jorde LB, Schiefenhovel W (2002) Affinities among Melanesians, Micronesians, and Polynesians: a neutral biparental genetic perspective. Hum Biol 74(3):413–30

Macintyre M, Foale S (2002) Politicised ecology: local responses to mining in Papua New Guinea. Resource Management in Asia-Pacific Program, Division of Pacific and Asian History, Research School of Pacific and Asian Studies, The Australian National University, Canberra

Maddison R, Mhurchu CN (2009) Global positioning system: a new opportunity in physical activity measurement. Int J Behav Nutr Phys Act 6(1):73

Matsui T (1989) Semidomestication: rethinking the origins of agriculture and Pastoralism [Semidomesutikeishon: Noko to Yuboku no Kigen Saikou]. Kaimeisha, Tokyo. (In Japanese)

McIntyre NE, Knowles-Yánez K, Hope D (2008) Urban ecology as an interdisciplinary field: differences in the use of "urban" between the social and natural sciences. In: Marzluff JM, Shulenberger E, Endlicher W, Alberti M, Bradley G, Ryan C, Simon U, ZumBrunnen C (eds) Urban ecology. Springer, Berlin, pp 49–65

Medin DL, Atran S (eds) (1999) Folkbiology. MIT Press, Cambridge

Mills M, Pressey RL, Ban NC, Foale S, Aswani S, Knight AT (2013) Understanding characteristics that define the feasibility of conservation actions in a common pool marine resource governance system. Conserv Lett 6:418–429

Miyauchi T (2001) Ju-min no seikatsu senrayku to komonzu: Soromon shoto no jirei kara [People's life strategies and the commons: a case in Solomon Islands]. In: Inoue M, Miyauchi T (eds) Komonzu no shakaigaku: Mori kawa umi no sigen kyodo kanri wo kangaeru [Sociology of the commons: thinking communal resources management of forest, river and sea]. Shin-yo-sha, Tokyo, pp 144–164. (In Japanese)

Moji K, Koyama H (1985) A time-saving spot-check method applied to a Sundanese peasant community in West Java. Man Cult Ocean 1:121–127

Morán EF (ed) (1990) The ecosystem approach in anthropology: from concept to practice. University of Michigan Press, Ann Arbor

Moran EF (1990) Ecosystem ecology in biology and anthropology: a critical assessment. In: Moran EF (ed) The ecosystem approach in anthropology: from concept to practice. University of Michigan Press, Ann Arbor, pp 3–40

Morren GEB (1986) The Miyanmin: human ecology of a Papua New Guinea society. UMI Research Press, Ann Arbor

Myers N, Mittermeier RA, Mittermeier CG, da Fonseca GAB, Kent J (2000) Biodiversity hotspots for conservation priorities. Nature 403(6772):853–858

Nagaoka T (1999) Hope Pukerane: a study of religious sites in Roviana, New Georgia, Solomon Islands. M.A. dissertation, Department of Anthropology, University of Auckland, Auckland

Nagaoka T (2000) Hope and other sacred places in Kokorapa. Booklet distributed at Nusa Roviana Village, New Georgia, Solomon Islands. University of Auckland, Auckland. http://westernsolomons.uib.no/docs/Nagaoka,%20Takuya/Nagaoka_2000.pdf

Nakano K (1978) An ecological study of swidden agriculture at a village in Northern Thailand. Southeast Asian Stud 16:411–446

Nakazawa M, Ishimori D (2004) Kyusokuna jinko-zouka no seiin to kiketsu [Cause and effect of rapid population increase]. In: Ohtsuka R (ed) Solomon Shoto: Saigo no Nettairin [Solomon Islands: the last frontier of tropical rain forest]. University of Tokyo Press, Tokyo, pp 55–81 (In Japanese)

Nakazawa M, Yamauchi T, Tanaka M, Ishimori D, Furusawa T, Midorikawa T, Ohtsuka R (2002) Community health assessment by urine dipstick screening in relation to the variety of lifestyles in the Solomon Islands. People Cult Ocean 18:35–44

NASA Landsat Program (2008) In: USGS (ed), Landsat ETM+ scene p086r066_L71089066_06620021201. USGS, Sioux Fall

Noda T, Takahata Y, Sato T, Ikoma H, Mochida H (1996) Physicochemical properties of starches from purple and orange fleshed sweet potato roots at two levels of fertilizer. Starch-Stärke 48(11–12):395–399

Nombo P, Leach J (2010) Reite plants: an ethnobotanical study in Tok Pisin and English. ANU E Press, Canberra

O'Sullivan JN, Asher CJ, Blamey FPC (1997) Nutrient disorders of sweet potato, vol 48. Australian Centre for International Agricultural Research, Canberra

Ohtsuka R (1983) Oriomo Papuans: ecology of sago–eaters in lowland Papua. University of Tokyo Press, Tokyo

Ohtsuka R (2004) Environmental preservation and community welfare under rural development projects: a comparative ethnoecological study in Asia and Oceania. In: Sasaki T (ed) Nature and human communities. Springer, Tokyo, pp 45–63

Ohtsuka R, Suzuki T (1990) Population ecology of human survival: biological studies of the Gidra in Papua New Guinea. University of Tokyo Press, Tokyo

Ohtsuka R, Inaoka T, Umezaki M, Nakada N, Abe T (1995) Long–term subsistence adaptation to the diversified Papua New Guinea environment human ecological assessments and prospects. Glob Environ Chang 5(4):347–353

Oppenheimer SJ, Richards M (2001) Polynesian origins: slow boat to Melanesia? Nature 410(6825):166–167
Perfecto I, Vandermeer J (2008) Biodiversity conservation in tropical agroecosystems. Ann N Y Acad Sci 1134(1):173–200
Perz SG (2003) Social determinants and land use correlates of agricultural technology adoption in a forest frontier: a case study in the Brazilian Amazon. Hum Ecol 31(1):133–165
Pikacha PG (2008) Wild west: rainforests of Western Solomon Islands. Melanesian Geo Publication, Honiara
Plotkin MJ (1995) The importance of ethnobotany for tropical forest conservation. In: Schultes RE, von Reis S (eds) Ethnobotany: evolution of a discipline. Dioscorides Press, Portland, pp 147–156
Pollock NJ (1992) These roots remain: food habits in islands of the central and eastern Pacific since western contact. Brigham Young University Institute for Polynesian Studies, Laie
Population Reference Bureau (2004) World population data sheet. Population Reference Bureau, Washington, DC
Prance GT, Balee W, Boom BM (1987) Quantitative ethnobotany and the case for conservation in Amazonia. Conserv Biol 1(4):296–310
Puwastien P, Burlingame B, Raroengwichit M, Sungpung P (2000) ASEAN food composition tables. Institute of Nutrition, Mahidol University, Bangkok
Racelis AE, Aswani S (2011) Hopes and disenchantments of religious community forestry in the western Solomon Islands. Ecol Environ Anthropol 6(1):26–38
Ragone D, Lorence DH, Flynn T (2001) History of plant introductions to Pohnpei, Micronesia and the role of the Pohnpei agriculture station. Econ Bot 55(2):290–324
Rappaport RA (1984) Pigs for the ancestors: ritual in the ecology of a New Guinea People. Yale University Press, New Haven
Read JL, Moseby K (2006) Vertebrates of Tetepare Island, Solomon Islands. Pac Sci 60(1):69–79
Redford KH (1991) The ecologically noble savage. Orion 9:24–29
Rennie SJ (1991) Subsistence agriculture versus cash cropping the social repercussions. J Rural Stud 7(1):5–9
Reyes-Garcia V, Godoy R, Vadez V, Apaza L, Byron E, Huanca T, Leonard WR, Pérez E, Wilkie D (2003) Ethnobotanical knowledge shared widely among Tsimane' Amerindians, Bolivia. Science 299(5613):1707
Reyes-Garcia V, Vadez V, Huanca T, Leonard WR, McDade T (2007) Economic development and local ecological knowledge: a deadlock? Quantitative research from a native Amazonian society. Hum Ecol 35(3):371–377
Rodriguez DA, Brown AL, Troped PJ (2005) Portable global positioning units to complement accelerometry-based physical activity monitors. Med Sci Sports Exerc 37(11):572
Romanucci-Ross L, Moerman DE, Tancredi LR (1997) The anthropology of medicine: from culture to method. Bergin and Garvey, Westport
Romer K, Renzaho AM (2007) Re-emerging conflict in the Solomon Islands? The underlying causes and triggers of the riots of April 2006. J Peace Confl Dev 10 (online publication). http://www.bradford.ac.uk/ssis/peace-conflict-and-development/issue-10/Conflict-in-Solomon-Islands-FINAL-EDIT.pdf. Accessed 5 Feb 2014
Romney AK (1999) Cultural consensus as a statistical model. Curr Anthropol 40(S1):93–115
Romney AK, Weller SC, Batchelder WH (1986) Culture as consensus: a theory of culture and informant accuracy. Am Anthropol 88(2):313–338
Roughan J (1990) The nongovernmental organization factor in development: a view from the Solomon Islands. Pac Stud 14(1):95–108
Sanderson S, Redford K (2004) The defence of conservation is not an attack on the poor. Oryx 38(2):146–147
SATOYAMA Initiative (2010) About: SATOYAMA Initiative. http://satoyama-initiative.org/. Accessed 5 Feb 2014
Schlosberg D (1999) Environmental justice and the new pluralism: the challenge of difference for environmentalism. Oxford University Press, Oxford

Schneider G (1998) Reinventing identities: redefining cultural concepts in the struggle between villagers in Munda, Roviana Lagoon, New Georgia Island, Solomon Islands, for the control of land. In: Wassmann J (ed) Pacific answers to Western Hegemony: cultural practices of identity construction. Berg, Oxford, pp 191–221

Scott JC (1976) The moral economy of the peasant: rebellion and subsistence in southeast Asia. Yale University Press, New Haven

Scott JC (1985) Weapons of the weak: everyday forms of peasant resistance. Yale University Press, New Haven

Sekine H (2001) Kaihatsu to Mukiau Hitobito: Soromon Shoto ni okeru Kaihatsu Gainen to Li-dashippu [People confronting development: concept of development and leadership in Solomon Islands]. Toyo Shuppan, Tokyo. (In Japanese)

Sheppard PJ, Walter R, Nagaoka T (2000) The archaeology of head–hunting in Roviana Lagoon, New Georgia. J Polyn Soc 109(1):9–37

Sillitoe P (1996) A place against time: land and environment in the Papua New Guinea highlands. Routledge, New York

Sillitoe P (1998) An ethnobotanical account of the vegetation vommunities of the Wola region, Southern Highlands Province, Papua New Guinea. J Ethnobiol 18:103–128

Smith EA, Wishnie M (2000) Conservation and subsistence in small-scale societies. Annu Rev Anthropol 29:493–524

Solomon Islands Government (1999) Medium term development strategy 1999–2001, volume 1: policy and strategy. Solomon Islands Government, Honiara

Solomon Islands Government (2000) Report on 1999 population and housing census: basic tables and census description. Solomon Islands Government, Honiara

Solomon Islands National Statistics Office (2011) Report on 2009 population and housing census: basic tables and census description. Solomon Islands Government, Honiara

Solomon Star News (2014a) Gold Ridge withdraws workers. Published on 15 August 2014

Solomon Star News (2014b) Mining battle ends, court rejects Sumitomo bid. Published on 25 September 2014

Statistics Office (1995) Solomon Islands 1993 statistical yearbook. Solomon Islands Government, Honiara

Statistics Office (1997) Village resources survey 1995/96, Statistical bulletin no. 10/97. Ministry of Finance, Honiara

Stephens D (1969) Changes in yields and fertilizer responses with continuous cropping in Uganda. Exp Agric 5(04):263–269

Suda K (1994) Methods and problems in time allocation studies. Anthropol Sci 102(1):13–22

Suga Y (2005) Safe control on resources in local societies: from the past toward the Future [Zaichi shakai ni okeru shigen wo meguru anzen-kanri]. In: Matsunaga S (ed) The environment: values of the safety [Kankyo: Anzen no Kachi towa]. Toshindo, Tokyo, pp 69–100. (In Japanese)

Takekawa D (2000) Hunting method and the ecological knowledge of Dolphins among the Fanalei villagers of Malaita, Solomon Islands. SPC Tradit Mar Resour Manag Knowl Inf Bull 12:3–11

Takeuchi K (2010) Rebuilding the relationship between people and nature: the Satoyama initiative. Ecol Res 25(5):891–897

Townsend PK (2000) Environmental anthropology: from pigs to policies. Waveland Press, Long Grove

Umezaki M, Ohtsuka R (2003) Adaptive strategies of highlands-origin migrant settlers in Port Moresby, Papua New Guinea. Hum Ecol 31(1):3–25

Umezaki M, Kuchikura Y, Yamauchi T, Ohtsuka R (2000) Impact of population pressure on food production: an analysis of land use change and subsistence pattern in the Tari Basin in Papua New Guinea Highlands. Hum Ecol 28:359–381

Umezaki M, Natsuhara K, Ohtsuka R (2001) Protein content and amino acid scores of sweet potatoes in Papua New Guinea highlands. Ecol Food Nutr 40:471–480

Umezaki M, Jung Su L, Kawakubo K, Jiang HW (2010) Kasokudokei to GPS wo kumiawaseta katsudou kyoudo no jikuukan patan hyouka shuhou [Evaluation of temporal-and spatial-patterns of physical activity: an application of accelerometer and global positioning system]. Descente Sports Sci 31:42–48. (In Japanese)

United Nations (1999) World urbanization prospects: the 1999 revision. United Nations, New York

USGS (2004) Shuttle radar topography mission, 3 arc second scene SRTM_ffB03_p089r066, filled finished B. U.o.M. Global Land Cover Facility. College Park

van Helden F (1998) Between cash and conviction: the social context of the Bismarck-Ramu integrated conservation and development project, National research institute monograph, 33. The National Research Institute, Port Moresby

van Oudenhoven FJW, Mijatović D, Eyzaguirre PB (2010) Bridging managed and natural landscapes: the role of traditional (agri) culture in maintaining the diversity and resilience of social–ecological systems. In: Bélair C, Ichikawa K, Wong B, Mulongoy K (eds) Sustainable use of biological diversity in socio-ecological production landscapes: secretariat of the convention on biological diversity, Montreal, Technical series no. 52. Secretariat of the Convention on Biological Diversity, Montreal, pp 8–21

Wagner J (2007) Conservation as development in Papua New Guinea: the view from Blue Mountain. Hum Organ 66:28–37

Wall JRD, Hansell JRF (1975) Land resources of the Solomon Islands volume 4: New Georgia group and the Russell Islands. Ministry of Overseas Development, London

Waterhouse JHL (1928) A Roviana and English dictionary: with English-Roviana index and list of natural history objects. Printed at the Melanesian Mission Press, Guadalcanal

West P (2005) Translation, value, and space: theorizing an ethnographic and engaged environmental anthropology. Am Anthropol 107(4):632–642

West P (2006) Conservation is our government now: the politics of ecology in Papua New Guinea. Duke University Press, Durham

Western Province Government (1991) Na Buka Vivinei Malivi pa Zinama Roviana (Roviana Custom Stories Book). Cultural Affairs Office, Gizo

White G, Lindstrom L (1997) Chiefs today: traditional Pacific leadership and the postcolonial state. Stanford University Press, Stanford

Whitmore TC (1966) Guide to the forests of the British Solomon Islands. Oxford University Press, Oxford

Whitmore TC (1998) An introduction to tropical rain forests, 2nd edn. Oxford University Press, Oxford

World Bank (2011) Data by country: Solomon Islands. World Bank. http://data.worldbank.org/country/solomon-islands. Accessed 24 Jan 2005

World Health Organization (2000) Obesity: preventing and managing the global epidemic, World Health Organization technical report series, 894. World Health Organization, Geneva

Yamauchi T (2007) Modernization, nutritional adaptability and health in Papua New Guinea highlanders and Solomon Islanders. In: Ohtsuka R, Ulijaszek SJ (eds) Health change in the Asia-Pacific region: biocultural and epidemiological approaches. Cambridge University Press, Cambridge, pp 101–126

Yen DE (1990) Environment, agriculture and the colonisation of the Pacific. In: Yen DE, Mummery JMJ (eds) Pacific production systems: approaches to economic prehistory. Australian National University, Canberra, pp 258–277

Zent S (2001) Acculturation and ethnobotanical knowledge loss among the Piaroa of Venezuela: demonstration of a quantitative method for the empirical study of traditional ecological knowledge change. In: On biocultural diversity: linking language, knowledge, and the environment. Smithsonian Institution Press, Washington, DC

Index

T. Furusawa, *Living with Biodiversity in an Island Ecosystem*, Ethnobiology,
DOI 10.1007/978-981-287-904-2

The manufacturer's authorised representative in the EU is Springer
Nature Customer Service Centre GmbH, Europaplatz 3, 69115 Heidelberg,
Germany. If you have any concerns regarding our products, please
contact ProductSafety@springernature.com

Printed and bound by CPI Group (UK) Ltd, Croydon, CR0 4YY
01/10/2026
02225901-0016